The Phoenix Project

# THE PHOENIX PROJECT
## An Energy Transition To
## Renewable Resources

Harry Braun

Published in 1990 in the United States of America by:
RESEARCH ANALYSTS
Scientific Publications & Productions
P.O. Box 62892
Phoenix, Arizona, 85082
(602) 969-3777

LIBRARY OF CONGRESS CATALOGING IN PUBLICATION DATA:
Braun, Harry W., III, 1948-
The Phoenix Project: An Energy Transition to Renewable Resources.
1. Energy. 2. Environment. 3. Hydrogen. 4. Solar.

Library of Congress Number: 89-60277

ISBN: 0-924600-00-4 (Hard Cover)
ISBN: 0-924600-01-2 (Soft Cover)

Printed and bound in the U.S.A.
By Publishers Press, Salt Lake City, Utah

*To Elva,*
*for her dedication to*
*critical thinking.*

# THE PHOENIX PROJECT

## Cover Credits

**Stellar Hydrogen.**
The fuel for the Sun and other stars.
Central portion of Eta Carinae Nebula.
*Courtesy of Kitt Peak National Observatory and
the National Optical Astronomy Observatories.*
See pages 92-95

**Nanobial Enzymes.**
Protein-based molecules
like these made a transition
to solar-hydrogen resources
roughly 3.5 billion years ago
when they developed chloro-
phyll-based energy conversion
systems. *Computer graphics
modeling and photography by
Arthur J. Olson, Ph.D.,
Research Institute of Scripps
Clinic, Copyright (C) 1985.*
See pages 95-101

**Arcology.**
A city without cars;
a rational and practical
alternative to urban sprawl.
*By architect Paolo Soleri.
Courtesy of Massachusetts
Institute of Technology (MIT).*
See pages 144-145

**Starship Hydrogen.**
A hydrogen-scoop
space biohabitat that
would have a comparable
land area of California
and Arizona combined.
*By Harry Braun. Courtesy
of Trans Energy Corp.*
See pages 139-147

**Ocean Thermal
Energy Conversion (OTEC)
System**. A solar-hydrogen
power plant that can work
24 hours a day, producing
fish food and fresh water
as a by-product. *Courtesy of
Lockheed Missiles & Space
Company.*
See pages 171-178

**Liquid Hydrogen-Fueled
National Aerospace Plane**.
Fleets of such aerospacecraft
will be necessary to construct
large space biohabitats
pictured above. *Courtesy of
McDonnell Douglas Corp.*
See pages 135-139

**Prospective Liquid Hydrogen
Cryogenic Tanker.** Transport ships
carrying liquid hydrogen are the
only true solution to eliminating
oil spills that occur from accidents
and routine tanker operations.
*Courtesy of General Dynamics Corp.*
See pages 212-214

**A point-focus concentrator
"dish" system.** (Manufactured
by McDonnell Douglas Corp.)
Such systems can be mass-
produced for large-scale
hydrogen production in automobile manufacturing facilities.
*Courtesy of Southern California
Edison Company.*
See pages 178-194

**Spaceship Earth.**
Will the mammals be able
to survive the increasing levels
of environmental contamination?
*Courtesy of National Aeronautics
and Space Administration (NASA).*
See pages 3-30

**A solar dish forest for large-scale hydrogen production.**
A prospective vision of a pollution-free future.
*By artist Pierre Mion
Courtesy of U.S. Naval Research Laboratory.*
See page 181

# Foreword

*The Phoenix Project* is an analysis and synthesis of some of the most significant information and/or ideas that could allow the U.S., in cooperation with other countries, to make an energy transition to renewable resources. It is a comprehensive review of both the problems and the promise that confronts the global human community.

There are many books and articles that have described the global environmental problems related to greenhouse gases, acid rain, stratospheric ozone depletion, and chemical contamination. *The Phoenix Project* is unique because it also defines well-documented and fundamental solutions to these global problems. The essence of the solutions involves an energy and industrial transition to renewable resources. Other key solutions related to economic stability, educational reform, the Cold War and the arms race it has spawned are also presented.

*The Phoenix Project* is an effort to light a candle in the darkness. It is my hope that this book will help to serve as a trigger mechanism for change, because humanity is as close to a kind of technological utopia as it is to oblivion. Because we are in the midst of an unprecedented information explosion, the most important question to ask is: *What is worth knowing?* This question underscores the necessity of sifting through the sea of data to extract the significant insights. There is not enough time to know everything, which now means that one of the most important educational functions is to prioritize information.

*The Phoenix Project* is a proposal that has evolved over many years of research. It is also a synthesis of key points and insights that have been published in scientific journals, newspapers, news magazines, books, or broadcast on network news affiliates over a period of many years. The individual bits of information are very much like the pieces to a puzzle, in that after a threshold of pieces have been put in place, one can begin to see "the big picture." Although many technical issues are reviewed, no prior technical or mathematical skills are required to understand the key points presented.

Perhaps most importantly, *The Phoenix Project* documents that it is possible to implement a global, renewable solar-hydrogen energy system that can operate in harmony with the Earth's biological life-support systems. Moreover, such an energy and industrial transition can begin immediately; it will provide an important economic basis upon which the arms race can be dramatically reduced; and the primary obstacles to implementing such an energy transition are not technical, but political.

HB

# Acknowledgments

I am especially indebted to the following individuals for their many special contributions and insights that have made this book possible. As has been said:

*"If we are able to see farther than some, it is only because we are standing on the shoulders of many who have gone before us."*

Albert Bartlett, Ph.D., Department of Physics, University of Colorado (Boulder, Colorado), for his many efforts to communicate the significance of exponential growth. I am also grateful to Mr. Renz Jennings (currently serving on the Arizona Corporation Commission) for making me aware of Dr. Bartlett's work.

Glendon Benson, Ph.D., President, Aker Industries (Oakland, California), for his extensive engineering research and development of advanced solar Stirling engine generator-sets (gensets) for hydrogen production. I am also indebted to James Castiglioni, a long time friend and research associate, for bringing to my attention the work of Dr. Benson and his colleagues.

Roger Billings, Director, American Association for Science (Independence, Missouri), for his life-long research and development of hydrogen energy production and end-use systems.

G. Daniel Brewer, P.E., Lockheed Corporation (Burbank, California), for the many contributions that he and his colleagues at Lockheed have made in the development of liquid hydrogen-fueled aircraft and related support systems.

Robert B. Liden, Trans Energy Corporation (Phoenix, Arizona), for his extensive engineering, economic and computer experience, particularly in the automotive and energy industries.

Cesar Marchetti, Ph.D., International Institute for Applied Systems Analysis (Austria), for his perspective and many insights on how the Earth's initial microbial "founding fathers" developed the first hydrogen energy system over three billion years ago.

Roy McAlister, P.E., Trans Energy Corporation (Phoenix, Arizona), for his comprehensive and detailed understanding of science and engineering, particularly in the areas of chemistry, materials, energy technologies and environmental systems.

Gerard O'Neill, Ph.D., for his book, *The High Frontier: Human Colonies in Space,* which provides a realistic engineering and economic analysis of large-scale space colonization. I am also indebted to Brian Quig for making me aware of the important work of Dr. O'Neill and his colleagues at the Space Studies Institute.

Anthony Pietsch, P.E., Garrett Turbine Engine Company (Phoenix, Arizona), who first directed my attention to point-focus-concentrator systems as the most efficient solar technology option.

Robert Ponds, P.E., Ford Aerospace & Communications Corporation (Newport Beach, California), for his insights into the similarities between manufacturing solar point-focus-concentrator systems and automobiles.

Dale Rycraft, P.E., Air Products & Chemicals (Allentown, Pennsylvania), for his insights that have resulted from his many years of experience in handling and transporting both gaseous and liquid hydrogen.

Walter F. Stewart, and his colleagues at the Los Alamos National Laboratory (Los Alamos, New Mexico), for their photographs and many insights into the use and handling of liquid hydrogen.

T. Nejat Veziroglu, Ph.D., and his colleagues who make up the International Association for Hydrogen Energy. Their extraordinary efforts in creating and managing an international peer-review hydrogen technical society have provided the foundation upon which an industrial transition to renewable resources can be built.

I am also deeply indebted to the other scientists and engineers that have prepared the papers that serve as the technical basis of this book, which have been published by *The American Institute of Aeronautics and Astronautics, American Journal of Physics, Chemtech, IEEE Spectrum, The International Journal of Hydrogen Energy, Mechanical Engineering, Nature, Oceanus, Science, Scientific American, the Society of Automotive Engineers.*

I am especially grateful to Elva O'Dell, without whose assistance this book would not have been possible. I am also indebted to Edwin (Pete) Dixon who helped me to get this book started -- and finished, and to Jay O'Malley, former Chairman of the Board of the O'Malley Companies (Phoenix, Arizona) who, in spite of the objections of many, allowed myself and our chief engineer, Tom Bird, the opportunity to conduct the extensive research that serves as the foundation of this book.

I also wish to thank Jim Castiglioni, Jerry Dellwo, Pete Dixon, Bob Liden, Roy and Kathy McAlister, Mary Moore, David Pizer, Brian Quig and Michael Swift, for their many long hours of discussions that resulted in countless insights and contributions. I am also deeply grateful to Elva O'Dell, Mary Moore, Pete Dixon and David Belskis for their proofreading assistance, and Charles Burke, James Thornton, John Smith and David DeVere for the refined artwork and technical drawings they provided. Finally, I wish to thank my mother Bertha, my sister Kathy and her husband Gary for their patience and understanding over the years, and my father who was always there when I needed him. - HB

## THE PHOENIX PROJECT

*Chapter 1: Utopia or Oblivion*
An overall discussion of the global environmental problems that now collectively threaten the habitability of the Earth, as well as a brief review of the solutions that are realistically available.

*Chapter 2: Exponential Icebergs*
The significance of exponential growth, and its relationship to global problems, solutions, information and education is presented.

*Chapter 3: Conventional Energy Considerations*
A review of the interrelationships between energy and environmental systems. Fossil fuel and nuclear energy options and resources are discussed.

*Chapter 4: Hydrogen*
The origin of hydrogen in the universe and in biological organisms is provided, along with a review of the use of hydrogen as a pollution-free "universal fuel" and solar energy storage medium. Safety, automotive and other applications are discussed, including the use of hydrogen in the space program.

*Chapter 5: Solar Technologies*
A review of the most viable solar technology options that have been developed, including photovoltaic cells, wind, ocean thermal and point-focus-concentrator "dish" systems.

*Chapter 6: Renewable Energy Resources*
An analysis of the physical resources (i.e., land, water, etc.) that will be required for the large-scale implementation of solar-hydrogen energy systems and resources.

*Chapter 7: Conclusions*
A review of what must be done to avoid the oblivion scenario. Controlled-environment agricultural systems are discussed, as well as the Bush Administration, and the hydrogen legislation that has been introduced in the U.S. Congress.

# THE PHOENIX PROJECT:
## An Energy Transition to Renewable Resources

Harry Braun

## Table of Contents

|  | Page |
|---|---|
| Chapter 1: Utopia or Oblivion | 3 |
|  |  |
| The Problem | 3 |
| The Hydrogen Economy | 6 |
| Resolvable Environmental Problems | 9 |
| Global Acid Deposition | 10 |
| Greenhouse Gases | 11 |
| The Carbon Cycles | 13 |
| CFC Molecules | 16 |
| Other Problems | 23 |
| Projections | 25 |
| Trends | 27 |
| Cold War Considerations | 28 |
| The Phoenix Project | 30 |
| Conclusions | 32 |
|  |  |
| Chapter 2:  Exponential Icebergs | 34 |
|  |  |
| Dr. Albert A. Bartlett | 36 |
| Background | 37 |
| The Concept of 11:59 | 41 |
| The Age of Exponentials | 43 |
| Positive Exponentials | 45 |
| The Information Explosion | 46 |
| Future Shock | 47 |

Strategies for Survival                          48
Education: Problems & Solutions                  49
Starting Young                                   51
Teaching Teachers                                52
The Power of Ideas                               53
Multiple Exponential Icebergs                    54
Conclusions                                      55


Chapter 3: Conventional Energy Considerations    56

Interrelationships                               57
Energy Basics                                    61
Crude Oil Reserves                               64
Exponential Expiration Time (EET)                66
The Oil Surplus                                  71
The Battery                                      73
Nuclear Power                                    75
Radioactivity                                    77
Decommissioning                                  78
Nuclear Waste Storage                            80
Questions of Safety                              82
Military vs. Civilian Reactors                   83
Advanced Reactor Considerations                  83
Nuclear Fusion                                   85
Cold Fusion                                      86
Nuclear Economics                                87
Conclusions                                      90


Chapter 4: Hydrogen                              92

In The Beginning                                 93
Primordial Hydrogen                              94
The Nanobes & Microbes                           95
Photosynthesis                                   100
The Water Former                                 103
The Universal Fuel                               104
Primary vs. Secondary Sources                    106
Hydrogen Storage Systems                         108
Hydrogen Hydrides                                108
Liquid Hydrogen                                  111

Liquid Hydrogen Disadvantages                116
The Hydrogen Engine                          119
Roger Billings                               121
Hydrogen Safety                              126
Hydrogen Explosions                          129
NASA                                         132
Hypersonic Aerospacecraft                    135
Space Habitats                               139
Starship Hydrogen                            141
Other Alternative Fuels                      148
Hydrogen Production                          149
Conclusions                                  152


Chapter 5: Solar Technologies                157

Photovoltaics                                159
Wind Energy Conversion Systems               161
Wind Power Disadvantages                     163
Vertical Vortex Generators                   167
Ocean Thermal (OTEC) Systems                 171
Dish Genset Systems                          178
Early Dish Stirling Systems                  184
Dish Genset Cost Estimates                   191
Line-Focus Systems                           194
Solar Economics                              196
Return On Investment (ROI)                   198
U.S. Energy Policy                           199
Conclusions                                  203


Chapter 6: Renewable Energy Resources        205

An Overview                                  205
Wind Energy Resources                        206
OTEC Resources                               207
Solar Genset Resources                       208
Implementation Lead-Times                    214
Water Considerations                         216
Water Options                                218
NAWAPA                                       220
Conclusions                                  226

Chapter 7: Conclusions                        227

    Solutions                                 228
    Controlled Environment Food Production
       Systems                             229
    The Bush Administration                    234
    Senate Bill 639                            239
    Government Regulation                       241
    Final Comments                             243


References                                    245

    Chapter 1: Utopia or Oblivion              245
    Chapter 2: Exponential Icebergs            248
    Chapter 3: Conventional Energy
       Considerations                      249
    Chapter 4: Hydrogen                        251
    Chapter 5: Solar Technologies              255
    Chapter 6: Renewable Energy Resources      257
    Chapter 7: Conclusions                     258


Index                                         259

About the Author                              272

## The Phoenix Project

Figure 1.1: The Earth.

Chapter 1

# UTOPIA OR OBLIVION

*The Problem*

In exploring the vast universe, astronomers and as-
trophysicists have found that there are more than a hundred billion galaxies. Within each galaxy are hundreds of
billions of stars, and orbiting most stars is a substantial
collection of planets, moons and asteroids. This means
there are theoretically billions of planets that are similar to
the Earth. At present, however, the Earth is the only
planet that is known to be capable of sustaining human
life. In spite of this remarkable fact, the surface of the
Earth may soon become uninhabitable for humans and
other mammals. This unfortunate event will not likely occur as a result of a nuclear war, but as a result of the
exponentially increasing chemical contamination of the
Earth's soil, water and atmosphere, along with the physical destruction of its ancient biological habitats. Such
environmental damage is somewhat like a nuclear war occurring in slow motion.

The needless death of the only planet known to support mammalian life by a seemingly mindless "bulldozer
culture" is no small consideration. It means that the era of
mammals, which has been evolving for over 100 million
years, may soon be coming to an end. For many species,
it is already too late. Indeed, there are now as many
species being exterminated as during the great mass ex-
tinction that ended the era of dinosaurs. This time, however, it is not the reptiles that are on the threshold of ex-
tinction, but the mammals.

The question is: *Are there going to be any survivors?*

There is an extraordinary irony that given the exponential advances in information, particularly in the areas of molecular biology and computer science, that it is reasonable to assume that disease and other molecular disorders could soon be eliminated. In addition, such molecular technology could eventually be directed to repair the environmental life-support systems that are presently being destroyed and/or contaminated on a global scale. Such developments, however, are predicated on the survival of a global civilization that allows tens of thousands of highly trained scientists to concentrate in highly specialized fields, and at this point, not only is the survival of civilization at risk, but the survival of humanity itself. The details and implications of the exponential nature of the global problems will be discussed in more detail in Chapter 2, but in the final analysis, it appears that the human species is simultaneously racing toward both oblivion and utopia.

It is the central thesis of this book that although the oblivion scenario is likely, it is neither necessary nor inevitable. Given the extensive global ecological damage that has already occurred, it is highly likely that regardless of what actions are taken now, a high price is going to be paid for the reckless disregard of the Earth's biological life support systems. However, if it is known that there are major problems ahead, it is possible to be prepared and thereby minimize the damage. Fortunately, the major obstacles to change are not technical, but political. The solution essentially involves the industrialized world making a transition to renewable energy and biological resources. But the most important consideration is that there is only a limited opportunity to take corrective action because the global environmental problems are worsening exponentially.

Although the global problems are in large part related to energy and industrial policies, there is no question that the problems are being compounded by the growth in human population and its resulting increased consumption of the Earth's nonrenewable resources. A major concern is the fact that primary environmental, energy, and economic problems that have been accumulating gradually over many years are not responsive to sudden changes. These global forces are like an automobile that is increasing its speed as it heads for a cliff. There is a point-of-no-return beyond which brakes and steering are useless. Once that threshold is passed, it will not matter if the driver applies the brakes with all his might and attempts to alter course. Even with all the wheels locked, the force of inertia will propel the automobile over the cliff into an uncontrollable disaster.

Because a large ship has a much greater mass and momentum than an automobile, it requires much longer lead-times to change course. Countries are much larger than ships, and global atmospheric and ecological systems are much larger than countries, which means the lead-times to take corrective action are much longer still. This is an important consideration whether one is dealing with the accumulation of pesticides in food chains, the accumulation of greenhouse gases in the atmosphere, or the depletion of the Earth's stratospheric ozone layer. This is the nature of the problems we face.

Perhaps the greatest tragedy of all is that the destruction of the Earth's life support systems does not need to happen in order for urban-industrial populations to maintain their automobiles and high-technology life-styles. Assuming it is not too late to take corrective actions, it is helpful to know what could make a difference in minimizing the impact of the many global environmental catastrophes that now seem inevitable.

One thing is clear. Many of the most serious global environmental and economic problems are related to the fact that the industrialized world is based on a fossil fuel energy system that does not factor in the environmental costs that are inevitable with such a system. Fossil fuels have a negative environmental impact from the point of exploration, recovery, transportation of the fuel, and in their final end-use when they are burned in the atmosphere. To resolve the problem, it will be necessary for the industrialized world to make a transition to a non-fossil fuel energy system that is capable of sustaining an increasing human urban-industrial population.

Such an energy option does exist. It has been referred to as the hydrogen energy system, but given the more comprehensive implications that are involved, many distinguished engineers and scientists simply refer to the new energy system as the "*hydrogen economy*."[1,2,3,4] *The Phoenix Project* documents a specific proposal that will allow the U.S., in cooperation with other countries, to make an energy, industrial and biological transition to renewable solar-hydrogen resources over the next several decades. Such a transition could insure global economic progress while allowing the industrialized world to begin living in harmony and balance with the Earth's biological life support systems.

"Phoenix" is an ancient Egyptian word that refers to a mythical bird that would never die. After having lived for several hundred years, this immortal bird would then consume itself in flames, and rise out of its own ashes to be reborn. In a similar context, it is technically possible for the industrialized world to rise from the ashes of the highly polluting fossil fuels (principally oil, coal and natural gas), and renew itself by developing renewable solar-hydrogen technologies and resources that are both inexhaustible and essentially pollution-free.

### The Hydrogen Economy

Hydrogen is not just another energy option like nuclear, solar, petroleum or coal. Rather, it is a "universal fuel" that can unite virtually all energy sources with all energy uses. Hydrogen is the most abundant element in the known universe; it is the primary fuel for the Sun and other stars; it can be manufactured from a wide range of sources, including water, and when hydrogen is burned, the water is reformed as water vapor, making it completely renewable. Hydrogen is essentially pollution-free when burned and is generally safer than gasoline in the event of an accident or collision. The transition to a hydrogen economy is also one of the key variables that makes an industrial transition to renewable solar resources technically and economically feasible. For this reason, Chapter 4 is devoted to hydrogen and its origin, production, and potential use as an energy medium.

There already is an extensive scientific and engineering brain trust, the *International Association for Hydrogen Energy* (Coral Gables, Florida), which has over 2,500 representatives from over 80 countries. These distinguished scientists and engineers hold international technical conferences every two years to review virtually every aspect of hydrogen production, transportation, storage, and end-use applications. Organizing and coordinating a fundamental industrial transition to renewable solar-hydrogen resources, however, is an enormous task, somewhat comparable to setting up the U.S. War Production Board in World War II. As a result, such a fundamental energy transition cannot happen without widespread public and political support.

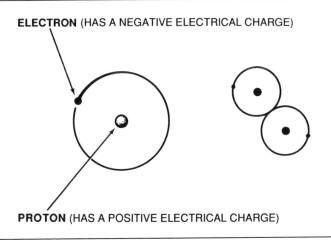

**ELECTRON** (HAS A NEGATIVE ELECTRICAL CHARGE)

**PROTON** (HAS A POSITIVE ELECTRICAL CHARGE)

Figure 1.2: The Hydrogen atom ($H_1$) & molecule ($H_2$).

There are many precedents for such an effort. During World War II the U.S. War Department organized the "Manhattan Project" which involved bringing together a large number of key scientists and engineers to focus upon the development of the atomic bomb. The American space program, the development of the railroads, or the interstate highway system are other examples of long-term "macro-engineering" projects that could not have happened without the successful cooperation of the government and private industry. A similar type of focused effort is going to be required if the human community is to make a successful transition to a renewable resource "stable-state" energy and economic system -- before the Earth's biological life support systems are irreparably damaged. After a certain point, it may be impossible to implement such a transition because the Earth's natural resources are being exponentially depleted at the same time that its biological life support systems are being exponentially destroyed and/or contaminated.

There already are reasons to be concerned. According to a recent report prepared by *World Watch* magazine, the U.S. has been typically producing about 300 million tons of grain annually. U.S. consumption is about 200 million tons and the rest has been exported. However, in 1987, U.S. grain production fell to 277 million tons, and in 1988, it fell to only 191 million tons because of the continuing drought. The only reason a global food crisis was avoided is because prior to 1987, there were record stocks of grain in storage[5]. Hopefully, the droughts will come to an end during the decade of the 1990's, but if the greenhouse global warming theories are correct, the droughts in general are not going to end -- *they are going to intensify.*

Civilization is a complex network of interdependent forces, and when major food-production and economic systems begin to disintegrate, social stability can disappear in a matter of weeks or days, if not hours. This is particularly true when one considers the psychological impact that could result when large numbers of people who are totally dependent on supermarkets, discover that they may soon be running out of food. Imagine the run on the local supermarkets by panic-stricken people who quickly realize that they have very little food stored. The stores would be emptied of food and other basic goods in a matter of hours. Most people have enough food to last several days, but if assistance is not forthcoming, or expected, the outlook for the future will be grim. There is already one-quarter of the human population that is without adequate food, but in a global systems collapse the vast majority of the Earth's human population could be lost in the chaos that could follow. If things are allowed to progress to this stage, an industrial transition to renewable resources will be as meaningless as trying to cash a check in a supermarket empty of food. The oblivion scenario will have transpired.

In order to cope with a potential collapse of existing global food production systems, it is necessary to be prepared for the multiple environmental dislocations that are, to a certain extent, inevitable. By acknowledging that there are many serious problems ahead and by planning accordingly, it is possible to minimize the impact of such problems. It will hopefully be possible to avoid the complete breakdown of the Earth's existing biological life support systems. While the global environmental problems are formidable, there are also many options to deal with the problems that will be documented in the subsequent chapters of this book.

*Resolvable Environmental Problems*

As an initial step in planning for constructive changes in the future, it is helpful to have a basic understanding of the global environmental problems that could be significantly resolved by making a transition to renewable solar-hydrogen energy systems. Some of the most serious problems that could be resolved by making such a transition include the following:

* Global acid deposition, which includes wet acid rain, fog and snow, as well as dry acid particles.
* Greenhouse effect gases, principally carbon dioxide, methane, and chlorofluorocarbon (CFC) molecules.
* Stratospheric ozone-destroying chlorine-based chemicals, principally CFCs, and methane molecules.
* Photochemical urban smog, benzene emissions, oil spills (including used automotive crankcase oil), leaks from gasoline and diesel fuel storage tanks, and the production of radioactive wastes.

*Global Acid Deposition*

Acid rain is the term that has received the most use in the media, but acid deposition (i.e., fallout) refers to a mixture of both wet and dry compounds of sulfur and nitrous oxides. The wet part of acid deposition can be in the form of rain, snow, hail, sleet, dew, frost or fog[6]. A solution's acidity is measured on the pH scale by its concentration of hydrogen ions. Because the pH value is a negative logarithm of this concentration, pH falls as acidity rises, and each full unit of change on the pH scale represents an exponential tenfold increase or decrease in acidity. The range is from 1 (highly acidic) to 14 (highly alkaline). A value of 7 is neutral.

Dr. Volker A. Mohnen, a professor of atmospheric science at the State University of New York at Albany, indicated in a *Scientific American* article that a lake becomes acidified when its pH falls below 6. Water collected near the base of clouds in the eastern U.S. during the summer now has a pH of about 3.6, but values as low as 2.6 have been recorded. The effects on forests have been devastating. According to Mohnen:

> *"Since 1980 many forests in the eastern U.S.
> and parts of Europe have suffered a drastic loss of
> vitality -- a loss that could not be linked to any of
> the familiar causes, such as insects, disease or di-
> rect poisoning by a specific air or water pollutant.
> The most dramatic reports have come from Ger-
> many, where scientists, stunned by the extent and
> speed of the decline, have called it Waldsterben, or
> forest death. Yet statistics for the U.S. are also
> unnerving."* [7]

Acid fog can be much worse than acid rain. Scientists at the California Institute of Technology and the University of California at Los Angeles have found acid fog to be *100 times* more acidic than acid rain[8, 9]. In addition, the dry acid particles and gases are also a serious problem. According to Dr. Robert W. Shaw, chief of chemical diagnostics and surface science at the U.S. Army Research Office, the dry deposition can be as destructive as the "wet" compounds. In a paper published in *Scientific American*, Shaw also indicated that acid deposition is generated primarily by the burning of fossil fuels:

> *"Studies tracing particle samples to their sources have helped to quash the notion that natural emissions from swamps, volcanoes or trees might be responsible for much of the acid fallout around the globe. It is now beyond question that even in rural areas acid deposition (both wet and dry) almost always stems from the activity of human beings: primarily the combustion of fuel for power, industry and transportation."* [10]

In excess of *40 million tons* of acid deposition is emitted into the atmosphere annually just from the U.S. Acid deposition not only severely affects biological systems, but also corrodes buildings, bridges, and the priceless ancient statues in Europe and the Middle East that have survived for centuries. While the vast majority of insects and microorganisms have been able to adapt to the pesticides that have thus far been developed, and even moderate exposure to radioactive isotopes, acid deposition is another matter. This is because the acids dissolve the very biochemical molecules that make life possible. Although acid deposition has not received as much media coverage as the stratospheric ozone hole or the greenhouse gases, it is every bit as serious a concern.


*Greenhouse Gases*

Greenhouse gases, which are principally made up of carbon dioxide, methane and the chlorofluorocarbons, trap the infrared heat from the Sun just like the glass in a greenhouse, thereby raising atmospheric temperatures. One of the many ways this "greenhouse effect" can directly impact weather conditions is by increasing droughts, which, in turn, contributes to the advancing of the deserts.

Globally, it has been estimated by Dr. Richard A. Houghton and Dr. George M. Woodwell, two senior scientists at the Woods Hole Research Center in Woods Hole,

Massachusetts, that the net atmospheric gain of carbon dioxide is about *3 billion tons* annually[11]. They also indicated in a *Scientific American* article, "Global Climate Change," that while the total amount of carbon dioxide in the atmosphere is only about *three one-hundredths of one percent,* in contrast to oxygen and nitrogen that together make up 99 percent of the atmosphere, oxygen and nitrogen are not greenhouse gases that absorb infrared radiation or radiant heat, whereas greenhouse gases like carbon dioxide, methane and chlorofluorocarbons do. *This means that in spite of their small concentrations, the greenhouse gases play a significant role in regulating the temperature of the Earth.*

Dr. James Hansen, who is the director of the National Aeronautics and Space Administration's (NASA) Institute for Space Studies and an expert on climate change, testified before the Senate Energy and Natural Resources Committee on June 23, 1988, that the Earth had been warmer in the first five months of 1988 than in any comparable period since measurements began 130 years ago. He attributed the increase in temperature to the long expected greenhouse effect, stating that:

> *"...it is time to stop waffling so much and say that the evidence is pretty strong that the greenhouse effect is here."* [12]

Woodwell, who is the director of the Woods Hole Research Center, has also provided congressional testimony that the widespread destruction of forests is significantly accelerating the global warming trend because the dying forests release the carbon dioxide they store in their organic matter. In addition, Dr. Syukuro Manabe and Dr. Richard Wetherald, of the U.S. National Oceanic and Atmospheric Administration's Fluid Dynamics Laboratory at Princeton University, have reported that carbon dioxide increases have also been implicated in drying out the soil in major agricultural areas of the U.S. Their study, published in the April 1986 issue of *Science,* indicates that the rising carbon dioxide levels could trigger a significant reduction in soil moisture in the grain belts of the U.S., Canada and Europe[13].

*The Carbon Cycles*

According to Dr. Wallace Broecker, Professor of Geochemistry at Columbia University and Director of the Center for Climate Research, the thermostat that determines the surface temperature of the Earth is the amount of carbon dioxide that is in the atmosphere, which in turn, is determined by the carbon cycle [14]. For hundreds of millions of years, the carbon cycle was maintained by rain washing the carbon dioxide from the atmosphere into the oceans, where the concentration is 50 times that of the atmosphere. The carbon dioxide is gradually deposited on the ocean floor, where it combines with calcium to form limestone. The limestone then moves with the geologic plates until it is returned to the molten interior of the Earth. Eventually, the carbon is returned to the atmosphere as carbon dioxide through volcanic eruptions. This geologic process results in an overall carbon cycle time of about 200 million years.

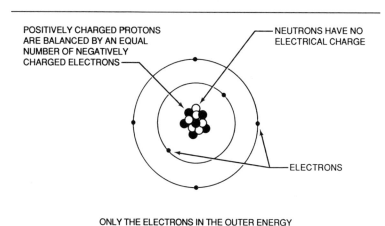

POSITIVELY CHARGED PROTONS ARE BALANCED BY AN EQUAL NUMBER OF NEGATIVELY CHARGED ELECTRONS

NEUTRONS HAVE NO ELECTRICAL CHARGE

ELECTRONS

ONLY THE ELECTRONS IN THE OUTER ENERGY LEVEL ARE INVOLVED IN CHEMICAL REACTIONS

Figure 1.3: The Carbon Atom.
Note that the carbon atom has six protons and six neutrons in its nucleus. Surrounding the nucleus in orbital paths are six electrons; two of the electrons are in the inner energy level, and the remaining four chemically active electrons are in the outer energy level.

Another major carbon cycle that has a relatively rapid cycle time involves living organisms. All life on the Earth is based on the assimilation of carbon and hydrogen. Green plants, for example, function by taking carbon dioxide out of the air, and by using the energy from the Sun, they combine it with water to form energy-rich carbohydrates. A by-product of this reaction is oxygen, which is released into the atmosphere. When the plant is consumed by other living organisms, energy is extracted from the carbohydrates, and the waste product of this cellular metabolism is carbon dioxide, which is again returned to the atmosphere[15].

If the plant material is not consumed and decomposes into one of the fossil fuels, the carbon is stored until such time as the fossil fuels are burned. Ice cores that have been cut from the polar ice caps have allowed scientists to analyze air that was trapped in the ice for thousands of years. As a result, it has been determined that at the peak of the Ice Age, some 20,000 years ago, the amount of carbon dioxide in the atmosphere was about 30 percent less than it is today[16].

Since the geologic carbon cycle takes hundreds of millions of years, Broecker believes that the relatively rapid carbon dioxide change may have been brought about by changes in the ocean population of plankton, the microscopic plants that are members of the vast and ancient algae family that anchors the aquatic food chain. Plankton absorb significant amounts of carbon dioxide, and when they die, they sink to the ocean floor, carrying the carbon dioxide with them. Thus, the greater the population of plankton, the more carbon dioxide they are able to take out of the atmosphere. When the plankton population declined during the last Ice Age, more carbon dioxide was retained in the atmosphere. This caused the temperature of the Earth to increase, which may have brought an end to the last Ice Age.

A more fundamental explanation has been put forth by Dr. James E. Lovelock, a distinguished British atmospheric chemist who believes the sum total of the Earth's microbial life forms, including the vast and ancient populations of bacteria, fungi and molds, make up a global living system. Lovelock refers to this global biological "superorganism" as Gaia, a Greek word meaning "Goddess of the Earth." He believes that this Gaian system has been actively regulating the climate of the Earth for over 3 billion years by engineering life forms that generate or absorb various atmospheric gases, including carbon dioxide, oxygen and methane.

Lovelock summarized his theory in two books, *Gaia: A New Look at Life on Earth*[17], and *The Ages of Gaia: A Biography of Our Living Earth*[18]. He points out that although the Sun has increased its heat output by roughly 30 percent since life began on the Earth, the surface temperature has remained relatively constant. Lovelock believes this highly unnatural phenomenon may have occurred because the sum total of the Earth's microbial biota, concentrated in the continental shelves of the oceans and great forests on land, were able to pump just enough carbon dioxide out of the atmosphere to keep the surface of the Earth from heating up along with its aging Sun. However, Lovelock believes that the Gaian force is reaching the end of its ability to regulate the Earth's climate. This is because the Sun is continuing to increase its heat output at the same time that carbon dioxide and other greenhouse gases are increasing exponentially due to the combustion of fossil fuels and the simultaneous destruction of the remaining rain forests.

The logical conclusion is that if the Gaian forces are no longer capable of pumping carbon dioxide out of the Earth's atmosphere, the existing climatic "stable-state" is soon going to be lost. This means the human population and other mammalian life forms may soon be extinguished in the rapidly changing climatic environment -- in the same way that the dinosaurs disappeared in a relatively brief period of time. The major groups of microbes will surely survive for a time by engineering new organisms that will be able to adapt to the more severe environment. But the surface of the Earth will be a vastly different place than it is now.

Regardless of whether a microbial superorganism exists or not, it is clear that the delicate balance of carbon dioxide and other gases in the Earth's atmosphere is a critical factor in allowing living organisms to survive and evolve. Ever since the industrial revolution, ever increasing amounts of carbon dioxide have been released into the atmosphere, principally as a result of deforestation and the human burning of fossil fuels. In the U.S., which has the highest per capita consumption of fossil fuels of any country, about six tons of carbon dioxide are released per person annually. Global carbon dioxide levels have increased by roughly 10 percent since 1900, and the rate is increasing exponentially[19].

Venus provides a graphic example of what happens when too much carbon dioxide accumulates in the atmosphere. Venus has no water to wash the carbon dioxide from its atmosphere, thus it has a surface temperature of about 850 degrees Fahrenheit. Mars, on the other hand, is an example of what happens when there is too little atmospheric carbon dioxide. Initially, Mars had oceans and a climate similar to that of the Earth. But because the continents did not drift on Mars, the carbon was trapped on the ocean floor. This resulted in increasingly cooler surface temperatures until the oceans finally froze[20].

## CFC Molecules

Although carbon dioxide is one of the most significant greenhouse gases, other serious offenders are the chlorine-based chlorofluorocarbon (CFC) molecules. CFCs are used as industrial solvents and in a wide range of products, including aerosol spray cans and foam packaging. But some of the most reactive CFCs are those that are used in the Freon of air conditioning systems. CFCs account for about 20 percent of the greenhouse gases, and also contribute to the acid rain problem, but their most serious effect is that they are one of the principal agents responsible for destroying the Earth's stratospheric ozone layer. Each year roughly *800,000 tons* of CFCs are released into the atmosphere globally[21]. Eventually, the CFCs will break down, thereby releasing chlorine, and *each chlorine atom can destroy over 100,000 stratospheric ozone molecules that protect living organisms on the Earth from the deadly wavelengths of ultraviolet radiation*[22,23,24].

Ozone is a relatively unstable molecule that is made up of three oxygen atoms, and for more than a billion years there has been a thin layer of stratospheric ozone floating about 15 miles above the Earth's surface. This ozone is continuously being created and destroyed by the Sun's ultraviolet radiation, but up until recently, it has remained in balance. This stratospheric ozone shield has been a critical factor in the evolution of life on the Earth, particularly with respect to humans and other animals who live on land. This is because the stratospheric ozone shield has been able to absorb the deadly short-wave ultraviolet radiation that is emitted by the Sun, while allowing the beneficial long-wave ultraviolet and visible radiation to pass through.

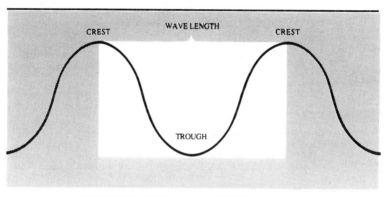

TRANSVERSE WAVE ——→ DIRECTION OF TRAVEL

Figure 1.4: Wavelength.

Solar radiation is made up of electromagnetic units that Albert Einstein referred to as photons. The energy level of a photon is dependent on its wavelength (refer to Figures 1.4 and 1.5), the shorter the wavelength, the higher the energy level of the photon. It is significant that as the wavelength decreases, the energy level of the photon increases exponentially. Biological organisms are made up of molecules that are, in turn, made up of a highly complex arrangement of atoms. Each atom contains a nucleus made up of protons (hydrogen nuclei) and/or neutrons, which are usually surrounded by a cloud of one or more orbiting electrons. The electrons carry a negative electrical charge, and they orbit the positively-charged nucleus somewhat like planets orbiting a star. As the electrons absorb photons of various wavelengths they become sufficiently energized to change their energy level, or break away from the nucleus of the atom altogether.

Electric Waves: 3,100 miles
(= 4,988 Kilometers)

Ultraviolet Radiation: 300 nanometers
(Note: A nanometer is one-billionth of a meter)

Figure 1.5: Wavelength and Energy Level.
As the wavelength decreases, its energy level increases.

The movement of electrons is the very basis of molecular chemistry, which includes everything from the synthesis of deoxyribonucleic acid (DNA) in our cells to making a cup of coffee. In the case of the high-energy short-wave ultraviolet radiation, the energy of the photon is so great that it can destroy the electrochemical structure of a molecule. Such molecular damage can cause healthy cells to mutate into cancer cells, or to die altogether. Up until recently, the Earth's protective ozone shield had been absorbing all of the dangerous ultraviolet photons (i.e., those below 290 nanometers). But as ozone-destroying CFC molecules began to be manufactured in larger and larger quantities in the 1950's and 1960's, they along with other ozone-destroying gases such as methane, began to drift up into the Earth's stratosphere where they were exposed to the high-energy, short wavelength ultraviolet radiation. This caused the CFCs to breakdown chemically, thereby releasing the chlorine atoms, which in turn destroy the ozone molecules.

Given that roughly *800,000 tons* of CFCs are released into the Earth's atmosphere annually, it should not be surprising that atmospheric scientists have begun to record serious depletions and "holes" in the Earth's stratospheric ozone layer. NASA scientists have been using sophisticated satellites and high-speed computers to monitor the stratospheric ozone levels since the early 1970's, but the first major hole in the ozone layer was discovered in 1981 not by NASA scientists, but by a small British research team, the *British Antarctic Survey*, using a single, relatively unsophisticated instrument.

At first, the British researchers, under the direction of Dr. J.C. Farman, thought the instrument they were using had been giving false readings. They were aware that NASA and other scientific teams had been using highly sophisticated instrumentation to measure stratospheric ozone levels, and they could not figure out why these other laboratories had not announced such a significant drop in the Antarctic ozone levels. What they did not know was that the NASA scientists had programmed their high-speed supercomputers to ignore any ozone readings that were off by more than a few percent, assuming them to be in error. Up to that point, none of the atmospheric scientists believed that drops of 20 or 30 percent in the stratospheric ozone shield were even possible. The measurements of Farman and his colleagues however, indicated that the Antarctic ozone levels had decreased by an alarming 50 percent in the last decade.

After carefully rechecking his data, Farman and his British colleagues published their results in 1985. Farman's initial results were essentially ignored by the media until the scientists at NASA went back and reexamined their computer data (which comes out at the rate of 200,000 data points per day) and confirmed Farman's observation that the massive stratospheric ozone hole was indeed there; it was the size of the continental U.S. and could be seen from Mars!

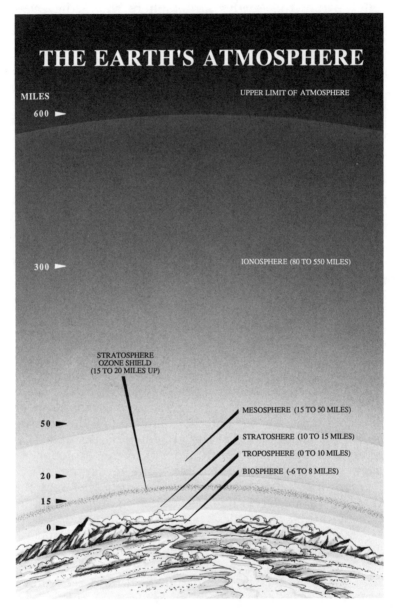

Figure 1.6: The Earth's Atmosphere [25].

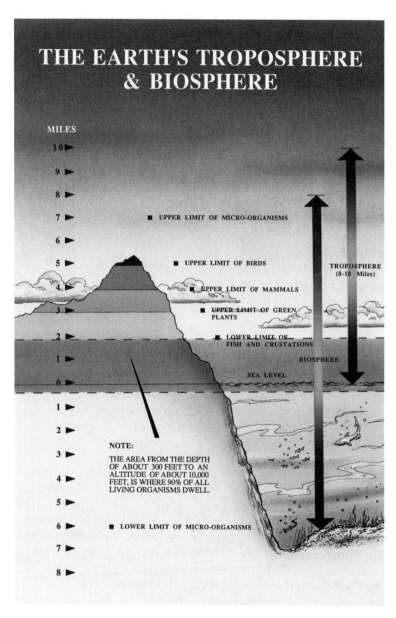

Figure 1.7: The Earth's Biosphere [26].

In addition to the large ozone hole in the Antarctic, overall global ozone readings over the past six years also indicated an overall decline of about 3 to 4 percent [27]. But there is another more ominous kind of instrumentation that is confirming the fears of ozone depletion. According to Dr. Darrel Rigel, a research physician from the New York University Medical Center, the rate of skin cancer in the U.S. is already increasing at a near epidemic pace -- outstripping predictions made as recently as 1982. Rigel, testifying before a hearing of the Energy and Commerce Health and Environmental Subcommittee on Ozone Depletion, stated that malignant melanoma, the deadliest form of skin cancer, has increased 83 percent from 1980 to 1987. It is now increasing faster than any other form of cancer with the exception of lung cancer in women. According to Rigel:

> *"Five years ago it was unusual to see people under 40 with skin cancer. Now, we often find it in people in their 20s."* [28]

Dr. Robert Watson, Director for upper-atmospheric research at NASA has made similar observations:

> *"There is now compelling observational evidence that the chemical composition of the atmosphere is changing at a rapid rate on a global scale."* [29]

The fact that exposure to the natural outdoor sunlight is becoming increasingly risky has ominous biological implications, because extensive scientific literature has documented that exposure to the natural outdoor sunlight can be as important to human health and productivity as nutrition and exercise [30,31,32]. Although the potential damage of chlorofluorocarbons to the Earth's ozone was widely reported in the press in the mid-1970's, the chlorofluorocarbon industries continued to disregard such warnings as "alarmist," which allowed them to maintain and even expand their multi-billion dollar business.

In 1977, the Environmental Protection Agency (EPA) banned the use of chlorofluorocarbons in aerosol spray cans, but the chlorofluorocarbon industries found other applications for their chemicals. As a result, the world-wide production of chlorofluorocarbons continued to increase at about 5 percent per year. An international treaty to reduce the amount of chlorofluorocarbons by 50 percent by the year 2010 was signed in 1988 by the major chlorofluorocarbon producing countries. But most atmospheric scientists believe a 50 percent reduction is inadequate. It is apparent that few people, including many experts, realize that the urgency of the situation is due to the massive inertial effects of these long-lived compounds. Although chlorofluorocarbon gases are often mentioned as the primary problem in stratospheric ozone depletion, other significant ozone killers include methane, which is often released into the atmosphere as a by-product of petroleum drilling, and nitrous oxides, which are a by-product of fossil fuel combustion.

*Other Problems*

Photochemical smog is a principal component of urban air pollution, and it includes dangerous increases in *tropospheric* ozone (i.e., that found in the lower atmosphere that extends from the Earth's surface to about 10 miles out). According to two Cornell University scientists, Dr. Peter Reich and Dr. Robert Amundson, tropospheric ozone can cause even more short-term damage to plants than acid rain [33]. This is to be distinguished from the *stratospheric* ozone levels that are being dramatically decreased. It is the stratospheric ozone (which is about 15 to 30 miles above the Earth) that protects plants and animals on the Earth by absorbing the deadly short wave radiation from the Sun.

There are other common, yet serious environmental contaminants that have not received much publicity. One example is the dumping of toxic crankcase oil that is used to lubricate the tens of millions of automotive engines. Although some of this crankcase oil is recycled, there are millions of gallons of it that are routinely dumped into back yards, vacant lots or landfills by motorists who either are unaware of the problem, or simply do not care.

There is also the unfortunate reality that there are literally millions of gallons of gasoline that leak into the ground annually from the storage tanks of neighborhood gas stations and other hydrocarbon fuel storage facilities. According to the EPA, there are an estimated 75,000 to 100,000 leaking fuel storage tanks that are releasing roughly *11 million gallons* of gasoline annually into the ground, and the number of faulty tanks is increasing[34]. Officials from the American Petroleum Institute have confirmed that the petroleum industry's estimates are about the same as the EPA estimates. This means that on average, there are at least 2,000 leaking fuel storage tanks in each state in the U.S.

Among the toxic substances in gasoline are benzene and ethylene dibromide, both of which are known to cause cancer. Because of the toxicity of gasoline, the EPA scientists indicated that such pollution poses a serious threat to the nation's underground water supplies. Mr. Jack Ravin, an EPA administrator for water, testified before a Senate hearing that one gallon of gasoline a day leaking into an underground water source is enough to contaminate the water supply of a community of 50,000 people. That means that if 11 million gallons are leaking into underground water supplies annually, over 150 million people in the U.S. could be affected. Although it is not known if humans and other mammals are able to adapt to such water pollution, the current cancer statistics are not encouraging. According to the American Cancer Society, more than 1 million Americans will learn they have cancer in 1989, bringing the total to 76 million. Roughly 500,000 Americans already die from cancer each year, making it the fourth leading cause of death[35].

If the chemical contamination problems were not bad enough, there are other serious global problems that critically compound the chemical contamination problems. Among these is a loss of forests (deforestation). As humans chop or burn down the remaining forests to make more room for themselves, they eliminate the trees that pump carbon dioxide out of the air, as well as attract clouds and moisture. In addition, deforestation accelerates the erosion of top soil and thereby contributes to the advancing of deserts. Desert regions, in turn, reflect the

Sun's heat creating a high pressure system that makes it more difficult for the cooler moist air to enter these areas. Thus, this deadly cycle feeds on itself[36].

*Projections*

Because of the complexity of global atmospheric and weather systems, no one can accurately predict how soon these problems will begin to affect the Earth's major food production systems. However, given the exponential nature of these problems, a major collapse of the Earth's life support systems could occur within the next 30 to 50 years, if not sooner. Indeed, if the droughts in the U.S. continue, serious food shortages could occur during the decade of the 1990's.

In a study funded by the U.S. Department of Energy, over 100 million weather records from 1861 to 1984 were analyzed to determine if any climate trends could be determined[37]. The study not only concluded that the planet was growing warmer, but that the trend is accelerating. The hottest three years of the 123 year period were 1980, 1981, and 1983, and five of the nine hottest years occurred after 1978. These data, coupled with the fact that a new record heat wave occurred again in 1988, prompted many atmospheric scientists to speculate that the dreaded greenhouse warming is beginning to affect global weather conditions. In the *Scientific American* article, "Global Climatic Change," Houghton and Woodwell summarized their findings as follows:

> *"The world is warming. Climatic zones are shifting. Glaciers are melting. Sea levels are rising. These are not hypothetical events from a science fiction movie; these changes and others are already taking place, and we expect them to accelerate over the next years as the amounts of carbon dioxide, methane and other trace gases accumulating in the atmosphere through human activities increase.*
>
> *"The warming, rapid now, may become even more rapid as a result of the warming itself, and it will continue into the indefinite future unless we*

*take deliberate steps to slow or stop it. Those steps are large and apparently difficult: a 50 percent reduction in the global consumption of fossil fuels, a halting of deforestation, a massive program of reforestation.*

*"There is little choice. A rapid and continuous warming will not only be destructive to agriculture but also lead to the widespread death of forest trees, uncertainty in water supplies, and flooding of coastal areas."* [38]

The article concludes by stating that:

*"These issues will persist throughout the next century and dominate major technical, scientific and political considerations into the indefinite future."* [39]

One thing is becoming increasingly clear, major climatic changes are already impacting many countries with severe droughts, record-setting hurricanes, forest fires and crop losses in the lower latitudes, and an abnormal increase in the number of icebergs that have been breaking away from the polar icecaps. While it is not possible at present to prove conclusively that the current global climatic changes are caused by the greenhouse gases, it is hard to imagine that the 3 billion tons of carbon dioxide that are being added to the Earth's atmosphere each year will not have a profound impact on the Earth's climate.

Our human ancestors lived at a time when concern about the natural environment essentially did not exist as a priority. If anything, the natural world was something to clear and conquer. Indeed, it was viewed as an inexhaustible resource. However, the awesome power of exponential growth has turned a once-pristine planet into what has increasingly become a contaminated industrial wasteland. In many respects, the struggle is about over. The bulldozer culture is now in the mopping-up stages. The remaining ancient wilderness habitats are now few and far between and before long, they too will be gone.

Despite this grim scenario, there is reason to believe that humanity is as close to utopia as it is to oblivion.

While the global environmental problems are awesome, so is the collective potential of the global scientific and engineering community. It must, however, be assumed that due to the massive damage already inflicted on the Earth's biosphere, a series of ecological disasters are now inevitable. This is an important point because if one is aware that certain food-production systems are going to fail, action can be taken to minimize the losses. But the hour is late and there is no time to waste.

*Trends*

The current trends are not promising. Essentially, the problem can be reduced to more and more people competing for fewer and fewer resources. The problem is global, and grows more serious with each passing day. For many of the millions of starving people in the world, it is already too late. Temporary food can be provided for the short-term, but it will do little but prolong the inevitable. In many cases, the additional food means more babies will survive only to starve later. The question is: *How much time do those who live in the industrialized world have before they too will be threatened by an economic and/or environmental systems collapse?*

While no one can accurately predict exactly when such economic and ecological disruptions will occur, it is clear that the problems are growing more serious with the passage of time. The situation is in many respects similar to an earthquake. Before an earthquake can occur, tremendous pressure must first be built up over long periods of time in the geologic plates. The longer the pressure increases, the more devastating the earthquake will be. In the same way, enormous pressures have been building for decades on the Earth's energy, environmental and economic resources, and sooner or later, their ability to sustain us will collapse.

However, unlike an earthquake, the environmental and economic systems collapse that will be inevitable if nothing is done, can -- and should be -- directly affected by human action and planning. It is not necessary to perish like so many lemmings rushing into the sea. There are alternative paths available, but the necessary changes need to be made before the impending economic and environmental dislocations foreclose our options to act rationally.

*Cold War Considerations*

The significance of the Cold War and the arms race it has fostered has been so profound that any overview of the global energy and environmental problems would be incomplete without some reference to it. This is not just because the U.S. and Soviet military forces have the capability to launch enough nuclear weapons to end civilization as it now exists. Although this is certainly a serious consideration, the probability of an all out, full-scale nuclear war occurring between the U.S. and the Soviet Union is remote, particularly under the leadership of Mikhail Gorbachev. Even if the weapons are never used, however, trillions of dollars have been expended in their research and development, and much of the expense has been paid for with deficit-financed dollars. Moreover, it is not just a question of money. The vast majority of the scientists and engineers in the U.S. and the Soviet Union have been involved in weapons research, in contrast to medical, energy or environmental research.

An even more ominous concern involves the fact that the fabrication of nuclear, chemical and biological weapons has resulted in the production of vast amounts of highly toxic chemical wastes. In the U.S., the Defense Department generates far more waste products than any private company, and even if the hundreds of billions of dollars that would be required to clean up such waste products were available, the technology to do so does not yet exist. Moreover, press reports from within the Soviet Union indicate that they too have similar unresolved toxic waste problems.

All of these serious problems are a direct result of an arms race that in the end, has proven to be technologically unwinnable. Indeed, many defense experts contend that the arms race has only made the world a more dangerous place in which to live. With submarine launched nuclear missiles, warning times have been reduced to minutes, resulting in a nuclear hair-trigger. Had a similar effort been undertaken in the development of renewable resource technologies and biotechnology, many of the fundamental economic and environmental problems that now threaten virtually every country's future would likely not exist.

A moratorium on the U.S.-Soviet arms race should be feasible because individual nations can always be expected to act in their own self-interest, and the Soviet Union needs to make a transition to renewable resources as much as the U.S. does. In addition, under the Soviet leadership of Mikhail Gorbachev, a new era of East-West cooperation has begun. Many press reports in much of the world are now suggesting that the Cold War is at last coming to an end, and terms like "the disarmament race" are now being used to characterize the weapon and troop reductions that are taking place in Europe. Equally significant changes are occurring within the Soviet Union itself, as well as many other countries, including Germany, Poland, and South Korea.

With such breathtaking changes taking place in so many parts of the world, the time has come for the human community to shift its resources from the arms race to the renewable resource technology race. *It is especially important to realize that such an energy and industrial transition could support the global economic systems that are now so dependent on arms production.*

If after several years of a moratorium on the arms race, the governments of the U.S. and the Soviet Union are still as mistrustful of each other as they have been in the past (which is highly unlikely), the arms race could always be renewed. Moreover, a moratorium on the arms race does not mean either country must disarm itself, and there is no question that both superpowers have more than enough conventional and nuclear weapons for defensive purposes. The primary objective of a moratorium is to stop making new generations of ever more sophisticated weapons of mass destruction. In this way, vast numbers of engineers and scientists -- as well as the financial resources -- can be redirected to an energy and industrial transition to renewable resources.

*The Phoenix Project*

The Phoenix Project is a specific plan of action that calls for a reordering of national priorities.  It details a well-documented strategy that could provide an economic structure to help the U.S. and the Soviet Union in particular, to declare a 15-year moratorium on the arms race. This is because the implementation of *The Phoenix Project* would allow both nations to redirect their financial and technical resources to making an energy transition to renewable technologies and resources.  Such a major reindustrialization effort would allow the defense-related industries to shift their emphasis from arms production, to producing the vast numbers of renewable energy technologies that will be required to displace the use of fossil fuels. Because such a reindustrialization effort will be primarily financed by the private sector rather than by tax dollars, it could help to provide balanced Federal budgets and economic prosperity for decades.

There is also the fact that the petroleum reserves in both the U.S. and the Soviet Union are expected to be seriously depleted in about 10 to 15 years.  Thus, the danger of a military confrontation over the remaining petroleum reserves in the Middle East (which are estimated to last for only another 40 or 50 years) will only increase with time. These realities mean the preservation of world peace and security may depend on whether or not the world is able to make a successful industrial transition to renewable energy technologies and resources.

But even if the Earth's fossil fuels were inexhaustible, it would still be necessary to make a transition to a global hydrogen energy system because of the environmental considerations.  The continued use of fossil fuels as the primary industrial combustion fuel is proving to have a profoundly negative impact on the Earth's climate and biological life support systems.  This means that implementing a transition to a clean and infinitely abundant supply of energy is one of the critical factors that could allow the interrelated global environmental, industrial, and economic infrastructures to be maintained.  However, in addition to reindustrializing the world around renewable resources, there must also be a profound change of attitude with regard to protecting the natural environment. Perhaps the most obvious consideration is that no product should be produced unless it can be ecologically recycled.

There also needs to be the recognition that a fair market international trading system is more important than having an unregulated free market that is only concerned with short-term profitability. Indeed, it is ironic that as long as American consumers are willing to allow products into the U.S. marketplace that have been produced in countries that have essentially slave-labor conditions, and virtually no environmental controls, the standard of living in the U.S. will continue to decline. Unfortunately, few Americans understand that the short-term "free market" thinking, and the bulldozer culture it has fostered, is one of the primary forces that has resulted in the destruction of the rain forests, the stratospheric ozone layer, and most of the other serious global environmental and economic problems that now threaten to make the surface of the Earth uninhabitable.

While short-term band-aid solutions that involve energy conservation, such as developing more efficient automobiles and appliances, will help to buy some time, ultimately such conservation measures will only prolong the inevitable. If new renewable sources of energy are not developed, conserving what is left has been described as somewhat like taking a slow walk down a dead-end street. In addition, in the U.S., the progress in reduced energy consumption has to a large extent been the result of closing down the energy-intensive factories that were once the backbone of American industry. It is no secret that if the U.S. imports its steel, aluminum, copper, consumer electronics and automobiles, it will use less energy. The devastating cost, however, has been to structurally displace millions of workers.

*Time is of the essence.* Because of the nature of exponential growth, if responsible action is not taken soon, the oblivion scenario may be closer than anyone suspects. Each generation throughout history has been faced with life-and-death problems and decisions, but due to the fact that many of the most serious energy and environmental problems are worsening exponentially, the magnitude of the problems now confronting the human community are unprecedented. Fortunately, the knowledge to deal with these problems is also increasing exponentially. It is for this reason that humanity is now on the threshold of utopia or oblivion. One can only hope that there is still sufficient time to take the constructive actions that are necessary to protect the biological life-support systems of the Earth and the miracle of civilization that it has allowed to evolve.

## Conclusions

Writing in *Scientific American*, William D. Ruckelshaus, the first administrator of the Environmental Protection Agency, stated that if humanity is to avoid the many environmental disasters that now appear to be inevitable, it will require "*a modification of society comparable in scale to only two other changes: the agricultural revolution of the late Neolithic and the Industrial Revolution of the past two centuries.*" While the first two revolutions were "*gradual, spontaneous and largely unconscious,*" Ruckelshaus points out that the Environmental Revolution "*will have to be a fully conscious operation, guided by the best foresight that science can provide*" -- an undertaking that would be absolutely unique in human history[40].

In the remaining chapters of this book, the specifics of how such a global energy and environmental revolution can be accomplished by making a transition to renewable energy resources will be described in some detail. If such a transition is successfully implemented, our reward will not be insignificant. For by allowing civilization to continue, we will ensure that the stunning developments in computer science, biotechnology and molecular medicine will continue. This, in turn, will ultimately lead to a biological transition to renewable resources, as molecular biologists continue to unlock the secrets of life and the nature of disease by more fully understanding the specific molecular mechanisms involved.

These molecular processes have been utilized by the vast and ancient microbial and nanobial organisms (such as proteins and riboneuclic acids) that are at the heart of metabolism, and of life itself. These biological "founding fathers," whose ancestors are more than 3 billion years old, will be discussed in more detail in Chapter 4, although their true significance to the life that exists on the Earth is only beginning to be fully understood.

In the next chapter, the critical nature of exponential growth and its relationship to education in general will be reviewed. Without a basic understanding of exponential growth, it is virtually impossible for one to understand the promise of the future, or the serious nature of the global environmental problems that now threaten the survival of our species. As such, there is no more important concept to comprehend than the nature of the "exponential age" in which we find ourselves.

Figure 2.1: *The Titanic.*

## Chapter 2

# Exponential Icebergs

The ocean liner *Titanic* was one of the major engineering milestones of its era. It was an elegant expression of what was technically possible, and it was thought to be unsinkable. Perhaps the most important lesson to be learned about the sinking of the *Titanic* is that it need not have happened. It has been reported that the captain of the *Titanic* knew full well there were icebergs in the area because his engineers on board had received numerous warnings from other ships who had reported seeing icebergs nearby. The other ships had either slowed or stopped, but not the *Titanic*, whose captain was apparently more concerned about setting an impressive transatlantic record. This reckless overconfidence was the first mistake.

The second mistake was that, because it was assumed that the *Titanic* was unsinkable, insufficient numbers of lifeboats were placed on board. As a result, even after the *Titanic* had hit the iceberg, and the passengers knew that the ship was going to sink in the icy waters, most of them could do nothing except wait for the inevitable. In the hours that remained, many of the doomed passengers changed into their best formal wear and tried to enjoy their last minutes as much as they could under the circumstances.

One might conclude from this experience that Mother Nature is very unforgiving of those who are unaware or who miscalculate.  The more important lesson, however, is that the sinking of the *Titanic* was a catastrophe that could have been avoided if prudent actions had been taken.  It is the central thesis of this book that the U.S., and in a larger sense, the Earth itself, are very analogous to the *Titanic*. While most people assume that our magnificent "ship of state" is unsinkable, there are a great many atmospheric and environmental scientists from around the world who have been issuing ominous warnings about multiple "icebergs" ahead.  It is important to heed the warnings, because the exponentially worsening environmental problems do not just threaten the U.S. -- they jeopardize the global life support systems of the entire planet.  The global contamination problem is like a nuclear war occurring in slow motion.  And because of the nature of exponential growth, the time to take corrective action is limited.  The passage of time is, therefore, a pressing concern.  The question is: *How much time do we have?*

To obtain an accurate "fix" on our position, it is necessary to understand the characteristics of exponential growth outlined in this chapter.  Indeed, one cannot appreciate the genuine sense of urgency until one understands the simple but generally unknown exponential concept of "*11:59.*"

*Dr. Albert A. Bartlett*

Dr. Albert Bartlett is a professor of physics at the University of Colorado (Boulder, Colorado) and past President of the American Association of Physics Teachers.  He has published numerous papers and articles on exponential (also called geometric) growth and its relationship to the consumption of energy and environmental resources. Bartlett has written that one of the greatest shortcomings of most elected officials is their inability to understand the exponential function and how it relates to the global energy and environmental problems.  He describes exponential arithmetic as probably the most important mathematics students will ever see.

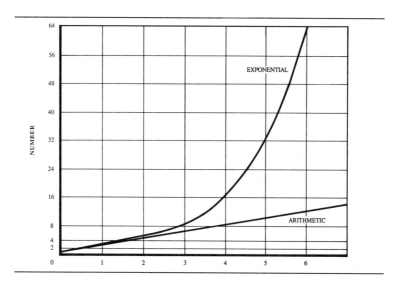

Figure 2.2: Arithmetic vs. Exponential Growth [1].

The importance of the exponential function is that it calculates steady growth. It is especially important to understand how a change in the sign of the exponent can make an enormous difference in the sum being calculated. What follows is a summary of Bartlett's key fundamentals of exponential growth that are described in his paper, "Forgotten Fundamentals of the Energy Crisis," published in the *American Journal of Physics* [2].

*Background*

When a quantity, such as the rate of consumption of a resource, is growing at a given percent per year, the growth is said to be "exponential." The important property of the growth is that the time required for the growing quantity to increase its size by a fixed fraction is constant. For example, a growth of 5 percent (a fixed fraction) per year (a constant time interval) is exponential. It follows that a constant time will be required for the growing quantity to double its size (increase by 100%). This time is called the doubling time, *T2*, and it is related to *P*, the percent growth per unit time by a very simple equation:

$$T2 = \frac{70}{P}$$

For example, a growth rate of 5 percent per year will result in the doubling of the size of the growing quantity in a time T2 = 70/5 = 14 years. But in two doubling times (28 years), the growing quantity will double twice (quadruple) in size. In three doubling times, its size will increase eightfold, and in four doubling times it will increase sixteen fold, etc. Exponential growth is characterized by doubling times, and it is easy to see how a few doublings can lead to enormous numbers.

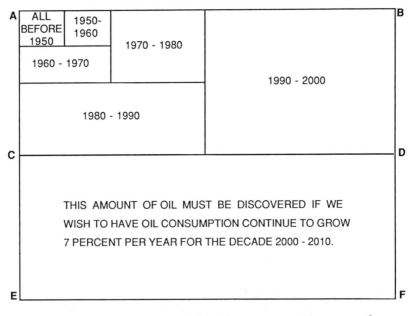

Figure 2.3: Exponential doubling times. The area of each rectangle represents the quantity of petroleum consumed in the labeled decade. The area of the rectangle ABCD represents the known world petroleum resource[3].

A simple but graphic example of exponential growth is to consider what happens if one penny is saved on the first day of the month, and each day thereafter, the amount is doubled. Few people realize that at the end of a thirty-one day month, over ten million, seven hundred and thirty thousand dollars would be saved. Observe what happens to the numbers over time in Table 2.1.

Table 2.1
Exponential Savings.

| Day | Amount ($) | Day | Amount ($) |
|-----|-----------|-----|-----------|
| 1 | 0.01 | 17 | 655.36 |
| 2 | 0.02 | 18 | 1,310.72 |
| 3 | 0.04 | 19 | 2,621.44 |
| 4 | 0.08 | 20 | 5,242.88 |
| 5 | 0.16 | 21 | 10,485.76 |
| 6 | 0.32 | 22 | 20,971.52 |
| 7 | 0.64 | 23 | 41,943.04 |
| 8 | 1.28 | 24 | 83,886.08 |
| 9 | 2.56 | 25 | 167,772.16 |
| 10 | 5.12 | 26 | 335,544.32 |
| 11 | 10.24 | 27 | 671,088.64 |
| 12 | 20.48 | 28 | 1,342,177.28 |
| 13 | 40.96 | 29 | 2,684,354.56 |
| 14 | 81.92 | 30 | 5,368,709.12 |
| 15 | 163.84 | 31 | 10,737,418.24 |
| 16 | 327.68 | | |

If a person's personal savings program were based on an arithmetic increase, the increases would occur at a fixed amount, rather than an amount that increases at an increasing rate. In Table 2.2, the amount saved each day is increased by a constant arithmetic rate of $0.01. Note that at the end of the 31-day month, only $0.31 has been accumulated. This is in contrast to the $10,737,418.24 that would result if the amounts were increased exponentially as they are in Table 2.1.

---

Table 2.2
Arithmetic Savings.

---

| Day | Amount ($) | Day | Amount ($) |
|-----|------------|-----|------------|
| 1 | 0.01 | 17 | 0.16 + .01 = 0.17 |
| 2 | 0.01 + .01 = 0.02 | 18 | 0.17 + .01 = 0.18 |
| 3 | 0.02 + .01 = 0.03 | 19 | 0.18 + .01 = 0.19 |
| 4 | 0.03 + .01 = 0.04 | 20 | 0.19 + .01 = 0.20 |
| 5 | 0.04 + .01 = 0.05 | 21 | 0.20 + .01 = 0.21 |
| 6 | 0.05 + .01 = 0.06 | 22 | 0.21 + .01 = 0.22 |
| 7 | 0.06 + .01 = 0.07 | 23 | 0.22 + .01 = 0.23 |
| 8 | 0.07 + .01 = 0.08 | 24 | 0.23 + .01 = 0.24 |
| 9 | 0.08 + .01 = 0.09 | 25 | 0.24 + .01 = 0.25 |
| 10 | 0.09 + .01 = 0.10 | 26 | 0.25 + .01 = 0.26 |
| 11 | 0.10 + .01 = 0.11 | 27 | 0.26 + .01 = 0.27 |
| 12 | 0.11 + .01 = 0.12 | 28 | 0.27 + .01 = 0.28 |
| 13 | 0.12 + .01 = 0.13 | 29 | 0.28 + .01 = 0.29 |
| 14 | 0.13 + .01 = 0.14 | 30 | 0.29 + .01 = 0.30 |
| 15 | 0.14 + .01 = 0.15 | 31 | 0.30 + .01 = 0.31 |
| 16 | 0.15 + .01 = 0.16 | | |

---

An important characteristic of exponential growth and doubling times is that the increase in any doubling is approximately equal to the sum of all the preceding growth. Note on day 10 of Table 2.1, for example, that the $5.12 saved is twice the amount of the total savings ($2.56) of the previous 9 days. Another important aspect of exponential growth is that even modest rates of growth can still eventually result in enormous consequences. For example, world production of crude oil increased only at a rate of about 7 percent per year from 1870 to 1970 but a 7 percent rate of growth means the doubling time is only 10 years. The danger of exponential growth is that when it first starts to grow, the level of growth seems insignificant, which leads to complacency. But after a few doubling times, even small amounts can increase into staggering quantities. Even small annual increases of 1 or 2 percent are significant.  Since 1900, the human population has been increasing at about 2 percent per year.  Yet, even at that modest level of growth, the human population will have increased from roughly 1.6 billion to 6 billion in the 100 years between 1900 and the year 2000.

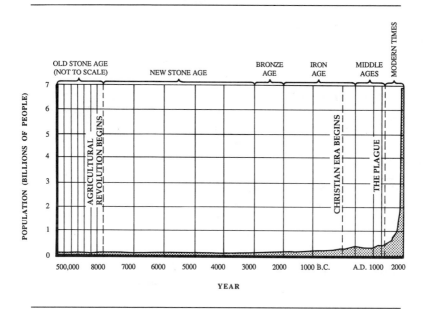

Figure 2.4: Human Population Growth[4].
It should be obvious to anyone looking at this graph why the existing human population growth is unsustainable.

*The Concept of 11:59*

In order to comprehend the sense of urgency with regard to the problems that now confront the human community, one needs to understand the exponential concept of "11:59."  As Bartlett points out, there are many different examples that can be used to explain the concept of 11:59. Some of the more common examples include lily pads growing in a pond, the number of people increasing on the Earth, or bacteria growing in a bottle.  Bacteria, for example, grow by division so that 1 bacterium becomes 2 and the 2 divide to give 4, etc. Assuming the division time is one minute, the bacteria will be growing exponentially with a "doubling time" of one minute. If one bacterium is put in an empty bottle at 11:00 in the morning, and it is observed that the bottle is full of bacteria at 12:00 noon, consider the following question:

When was the bottle half-full?
*Answer: 11:59*

If you were one of the bacteria in the bottle, at what point would you first realize that you were running out of space and therefore "resources?" Consider that at 11:55, the bottle is only 3 percent filled, which leaves 97 percent open space -- just waiting for "development." As Bartlett suggests, suppose that at 11:58, some farsighted bacteria realize that they are rapidly running out of resources, and with a great expenditure of effort they launch a search for new bottles. Let us assume that as a result of their efforts, the bacteria are fortunate enough to discover three new empty bottles. Great sighs of relief would be expected to come from all the worried bacteria. But even with the total space resource quadrupled, the exponential growth of the bacteria could only be maintained for another *two minutes!* Table 2.3 summarizes what happens to the numbers over time:

---

Table 2.3
Bacterial Exponential Growth.

---

| 11:55 a.m. | Bottle #1 is  3% full. |
| 11:58 a.m. | Bottle #1 is 25% full. |
| 11:59 a.m. | Bottle #1 is 50% full. |
| 12:00 noon | Bottle #1 is 100%  full. |
| 12:01 p.m. | Bottles 1 and 2 are  full. |
| 12:02 p.m. | All 4 bottles are full. |

---

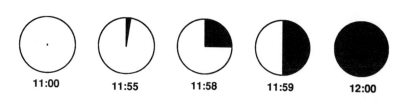

Figure 2.5: 11:59.

*The Age of Exponentials*

Mother Nature, which is essentially the laws of physics and chemistry, has consistently been shown to be very unforgiving to those who violate the laws by exceeding the physical limits of the ecosystem in which they live. Progress is generally viewed as a good thing, but given the nature of exponential growth, it appears that what the human community is progressing toward with ever-increasing speed are numerous exponential icebergs. Key problem areas (icebergs) that have been worsening exponentially for many years, if not decades, include the following:

> Human population growth
> National debt
> Acid deposition
> Carbon dioxide accumulations in the atmosphere
> Stratospheric ozone depletion
> Production of toxic chemicals
> Destruction of forests (Deforestation)
> Advancing of the deserts (Desertification)
> Consumption of nonrenewable fossil fuels
> Cancer and heart disease
> AIDS virus afflictions
> Poverty
> Destruction of marine and wildlife habitats

*Resources*

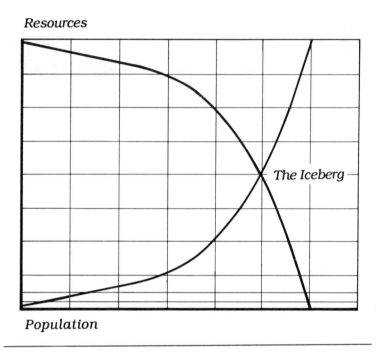

The Iceberg

*Population*

Figure 2.6: The Iceberg.

We are living at a time when the exponential increases in a large number of key environmental areas are (to borrow a term from the nuclear physicists) going critical. That is to say, in many areas, after a certain point in time, it will no longer be possible to alter the outcome. The primary concern is what happens when the resource capabilities of the Earth are consumed by the exponential human population growth. The answer can be referred to as a "systems collapse," whereby population levels are catastrophically reduced. A more common term is "die-off," but whatever one calls it, it is serious business.

The long-term impact of the exponential consumption of conventional fossil fuel and nuclear energy resources will be discussed in more detail in the next chapter. However, the key point to understand is that oil accounts for the vast majority of energy consumption in the U.S. (as well as most of the rest of the world) and the estimates from the American Petroleum Institute and the U.S. Geological Survey indicate that with current rates of consumption, the U.S. will exhaust its current oil reserves (which represent only about 4 percent of the remaining global reserves) in about 10 years. While additional oil will surely be discovered, it will be increasingly expensive to extract because it will be deeper and harder to find.

Even if the supply of oil and other fossil fuels were unlimited, their continued use as a combustion fuel could have a catastrophic impact on the Earth's ecological life support systems. As a result, the U.S. and other industrialized countries should be moving with wartime speed to make a transition to renewable resources. In spite of the incalculable global issues involved, however, no coherent strategy has yet evolved.

*Positive Exponentials*

It is important to realize that exponential growth does not have to be negative. It depends on what is growing exponentially. A savings account that is growing (i.e., compounding) exponentially is obviously a highly desirable asset to own. As financial planners are fond of telling their clients, even a one or two percent increase in the annual rate of interest can make an enormous difference in one's savings over a period of ten or twenty years.

Although the exponential growth in knowledge and information is much more difficult to quantify, there is no question that it is increasing at a staggering rate. This is particularly the case in the technical areas of engineering, computer science, molecular biology, and medicine. It is only because of these explosive developments in information and technology that it seems reasonable to expect that molecular medicine will soon usher in an age where molecular disorders and the diseases they cause will essentially be eliminated. With such fundamental changes in human life span potential, a future of unlimited possibilities awaits those who will be able to take advantage of such biotechnologies.

Such options are obviously based, however, on an assumption that the human community will be able to survive the many serious global environmental problems that continue to worsen exponentially.  Never before has humanity been at the crossroads of such awesome opportunities -- or problems -- as the exponential forces of life and death are simultaneously evolving and racing toward some ominous conclusion.  What needs to be understood by every thinking person is that the decisions that are made now could well make the critical difference in the ultimate outcome.  This places an extraordinary responsibility on the print and electronic news media to communicate the fundamentals of both the problems and solutions to the general public, while there is still time to take corrective actions.

### The Information Explosion

Of all of the phenomena that are increasing exponentially, none is more significant, or more promising, than the increases in knowledge and information.  Dr. Carl Sagan, professor of astronomy and space sciences at Cornell University, summarized the unprecedented information explosion in his book *Cosmos*:

> *"The great libraries of the world contain millions of volumes.  If I finish a book a week, I will read only a few thousand books in my lifetime, about a tenth of a percent of the contents of the greatest libraries of our time.  The trick is to know which books to read."* [5]

The rates at which knowledge, and its direct spinoff, technology, have been increasing is probably incalculable. Even to try and stay current in one specific area of medicine or engineering is exceedingly difficult, if not outright impossible. As a result, the most significant question to be asked in this "age of exponentials" with its seemingly endless sea of information is: *what is worth knowing?* Ultimately, it is a question of priorities, and priorities to a large extent depend on one's awareness of what is happening in one's immediate environment. If one's house is burning down, priorities are easy to establish because of the immediate and obvious threat. But problems dealing with acid deposition, radioactive waste, or ground water contamination are complex and the threat is not immediately observable. Effective solutions, if they exist at all, are usually very expensive. This adds an additional dimension to the problem. Specifically, who is going to pay to resolve the problem?

Such complex issues are confusing even for highly trained specialists -- much less members of the general public. This no doubt explains why such large numbers of people feel helpless to do anything. Because they cannot even begin to cope with the tide of events sweeping over them, the understandable response is to become apathetic and "tune out." Given this perspective, it is easy to understand why depression, mental illness and drug addiction problems are so widespread, as increasing numbers of people find they are unable to cope with a high-pressure, high-technology society that continues to change at a very rapid exponential rate.

*Future Shock*

The concept of "future shock" was first described in 1965 in a *Horizon* magazine article written by Alvin Toffler. Toffler later elaborated on the concept in his book, *Future Shock*, which explained that it refers to the shattering stress and disorientation that can be induced in individuals by subjecting them to too much change in too short a time. In some societies, centuries can pass with virtually no significant changes taking place. But in contemporary urban-industrial societies, the exponential acceleration of change has, in and of itself, become "an elemental force," and increasing numbers of people are simply unable to emotionally adapt to it. As Toffler wrote:

> *"Whether we examine distances traveled, altitudes reached, minerals mined, or explosive power harnessed, the same accelerative trend is obvious. The pattern, here and in a thousand other statistical*

*series, is absolutely clear and unmistakable.   Millennia or centuries go by, and then in our own times, a sudden bursting of the limits, a fantastic spurt upward."* [6]

Toffler went on to assert that this acceleration of change has now reached a point where it can no longer, by any stretch of the imagination, be regarded as "normal." Yet, the rate of change continues to increase at an ever increasing rate.  What becomes increasingly obvious is that technology seems to have a life of its own.

It is important to remember, however, that the information explosion and the rapid technological change it has stimulated is as much a part of the solution as it is the problem.  Indeed, there are many people who seem to be thriving in this era of rapid change.  Such individuals are not concerned about the future arriving too soon, they wish it would hurry up.  Of course, such people usually have the general assumption that things are going to get better in the future, as indeed they may.  But while it is important to keep a positive outlook, it is equally important not to forget about the problems that are compounding exponentially because if we do not deal with them, they are going to deal with us.

*Strategies for Survival*

1.  It is important to be part of the solution and not part of the problem.  Not being aware of the problems will not make them go away.  Rather, the exponential problems must be dealt with while there is still time to take corrective action.

2.  Understand that the situation is not hopeless. Humanity is as close to utopia as it is to oblivion. This is no time to give up.  While it is true that the problems are unprecedented, so are the technical resources that are now available to solve these problems -- *if we choose to use them for such purposes.*

3. Since 1945, the U.S. and the Soviet Union have allowed much of their human, financial and technical resources to be used in an arms race that, from a technology perspective, has been shown to be unwinnable. This is a political problem, and political problems require political solutions.

What is required is a reordering of the national priorities of not only the U.S. and the Soviet Union, but of all countries. Before that can happen, however, the global human community must be made aware of the realistic options that are available. The problem is that most of the options involve technical issues and questions that place an increasing emphasis on the education of not only the young, but *everyone*.

*Education: Problems and Solutions*

Education is the foundation of political action. But given the nature of the information explosion, the role of education takes on a new meaning. It is no longer just a question of absorbing information, because one can only assimilate a small fraction of that which is available. As a result, the most important educational decisions an individual can now make are those which revolve around the question of what is worth knowing?

There have been numerous news accounts of reports by various experts in education who have been critical of the U.S. educational system, particularly in the primary and secondary levels. One such report was prepared by Dr. Glenn T. Seaborg, a Nobel Laureate who has been a technical advisor to many presidents. Seaborg is a former member of the Manhattan Project during World War II, and he served as National Chairman for the development of a high school chemistry course. He is currently an active researcher at the University of California, and in 1983, he was appointed as a member of a special national commission that was to investigate the state of American education. After many months of research, the commission concluded the following:

*"If an unfriendly power had attempted to impose upon the United States the mediocre educational performance that exists today, we might have viewed it as an act of war."*

In the 1988 Public Broadcast System (PBS) television program *Innovation*, narrator Jim Hartz interviewed Seaborg who indicated that the situation is not getting any better, primarily because students are not being "turned on" by teachers that are often not interested in, or well-trained in, science and mathematics. When asked, "If it takes inspirational teachers to bring students along, and we are not producing very many, what's the answer?" Seaborg replied, "That's a very difficult problem." He went on to explain that it is a matter of restoring elementary and high school teachers to the status that they used to enjoy -- and that they still *do* enjoy in countries such as Japan and Germany.

When the U.S. educational system is compared to the one in Japan, there are many sharp contrasts. Per-capita student expenditures in Japan are reported to be less than those of U.S. students, yet according to a 1988 CBS *60 Minutes* broadcast titled "Head of the Class," the average Japanese student attends school 240 days per year, compared to 180 days for the average American student. In addition, the average American student's school day starts at 8:00 a.m. and ends at about 3:00 p.m, whereas the average Japanese student begins at about 7:00 a.m. and continues until midnight. While there are many Americans who would understandably be opposed to such a demanding system, there is the hard reality that it is not just the Japanese educational system that the U.S. is lagging behind. America's educational system compares unfavorably to those in virtually all of the industrialized countries and even many Third World countries.

Perhaps the most obvious solution would be to have the U.S. educational system adopt comparable academic standards to those in Japan and Europe for each grade level. If the students do not pass a course with at least an 80 or 90 percent proficiency, they are simply not allowed to progress to the next level. The concept of "social passing" should be eliminated. After all, it is what you know, and not how old you are that is important. Part of the problem, however, is a pervasive attitude of many American students (and adults) that can be reduced to the reality that they "live to play." This attitude begins early in life, as children in contemporary American culture are taught to play instead of using their time in a more constructive manner during their most formative and impressionable young years.

*Starting Young*

It is important to realize that mastering a scientific discipline initially involves learning to speak its language, and it has been observed that young children are able to absorb information or acquire languages much easier than their adult counterparts. If this is the case, it is especially important not to let young children idle away a critically important part of their formative years. A generation ago, the typical American child who grew up on a farm was usually given important responsibilities at an early age. This is still true in much of the Third World. In sharp contrast, however, U.S. children are rarely exposed to serious responsibilities until after high school, or in many cases, college.

Consider how much time many high school students invest in playing football or baseball. If the same effort were invested in mathematics, chemistry, biology, or physics, American students would be well on the way to being competitive with their Japanese or European counterparts. This is not to suggest that students should not participate in athletic events, but some sort of perspective needs to be maintained. It is advisable to reexamine the notion of encouraging students to spend thousands of hours developing short-term athletic skills, rather than long-term intellectual skills. Not only are there many serious accidents that occur in some athletic events, but the question needs to be asked: *Who cares how far somebody can jump, or how fast they can run with a ball?* If a comparable expenditure of effort were devoted to intellectual skills, one would have knowledge that would be useful throughout one's life. It is also important for each individual to observe how they are using their "free" time. This free time is an incredibly valuable resource that many people just idle away. If it is used wisely, it compounds over the years like interest in the bank. This, however, is a lesson that is rarely taught in most classrooms.

## Teaching Teachers

While the attitude of students and parents is certainly an important factor in one's educational development, there is no question that the serious problem raised by Seaborg is of paramount concern; namely, that there are few primary or secondary teachers that are equipped to teach technical subjects at all, much less in a creative and inspiring way. There is, however, a way out of this maze. While there may not be many teachers who are particularly gifted in teaching mathematics or science, there is no question that there are more than a few. This being the case, and given the miracle of video tape, it is certainly possible to have production quality films and/or video tapes produced and shipped to every school in the country, on every major area of scientific investigation.

Such video programs, which could and should be funded by the Federal Government, would allow the poorest inner-city or rural school children to have access to the very best instructors that exist in the country -- or the world for that matter. In addition, the programs could be made in such a way that the teacher is not just standing in front of a classroom lecturing. Rather, "Star Wars" quality modeling and computer animation could be incorporated to make learning and education truly entertaining, which it should be anyway. There is a special bonus in this approach. As the teachers are showing the high-quality video tapes to their students, the teachers themselves will soon be able to master the material presented, and thereby have their own education significantly upgraded while "on the job." As any teacher knows, the best way to master any subject, is to teach it.

Another important bonus to this approach is that if a student had trouble understanding all of the material during class (or perhaps the student was out ill), he or she can take the video home, and watch it in the evening as many times as it takes. This not only provides for individualized learning, but perhaps the family can be present so learning can become an enjoyable family affair. This will also provide parents with an opportunity to gain a better understanding of what their children are being taught in school, and allow the parents to upgrade their education as well.

There have already been many high-quality documentaries developed by the Public Broadcasting System network, including the *Nova* and *Frontline* series. What is tragic is that so few people watch these consistently outstanding productions. This reinforces the need to consider carefully how young people, as well as adults, use the limited amount of leisure time that they have available. It appears that a great many people spend their free time reading low-grade fiction (assuming they read at all), or watching televised entertainment programs, such as soap operas or television sitcoms, in contrast to the documentaries that could help them make more informed judgments. Given the serious nature of the problems that are continuing to compound exponentially, the net result of this unawareness could well be catastrophic.

### The Power of Ideas

There is a phrase often used by nutritional consultants that states "you are what you eat." In an intellectual context, one could say that *you are what you think*. This is primarily because individual thoughts and ideas can exert a powerful influence over how an individual feels. For example, if one thinks about unpleasant thoughts, such as the death of a family member or close friend, it is easy for one to become depressed and despondent. In contrast, there are other pleasant or happy thoughts and ideas that can instill confidence and/or provide one with a positive self-image and attitude.

It has been said that individual thoughts and ideas are in many respects like living organisms, in that they tend to reproduce their own kind. If, for example, negative or idle thoughts are allowed to dwell in one's mind, these are the kinds of thoughts that will tend to reproduce and multiply. On the other hand, if one makes an effort to try to concentrate on constructive and positive thoughts, those will be the thoughts that will tend to grow and multiply. To a certain extent, each person can and should determine content and the quality of the ideas and thoughts that occupy their conscious mind. Simply put: Garbage in - garbage out.

## Multiple Exponential Problems

It is unfortunate that the nature of exponential growth is not understood by most people. That will not, however, alter its inevitable result. Moreover, the dilemma is further compounded by the fact that the U.S., and the rest of the world, is not just threatened by one exponential problem, but many. It is critical that every thinking and voting person understands both the nature of the problems -- as well as the solutions -- that are presently available. In this regard, the process of communication and education is absolutely essential if there is to be any chance of implementing the necessary changes in time.

We can only hope that it is not already too late to avoid the more serious environmental and economic problems that not only threaten human civilization, but the Earth's biological life-support systems that have been evolving for billions of years. Those who attempt to detach themselves from the actuality that threatens the entire planet are a major part of the problem. This detachment is a delusion, because the Earth's biological life support systems are swiftly approaching the crossing of the exponential curves whereby the ascending line of consumption will intersect the descending line of available resources. If such a progression is allowed to continue, in the end, there will be no place to hide.

The urgency of the situation is due to the massive inertial effects that result from exponential growth. Even if the captain of the *Titanic* had found out about the iceberg before the point of impact (which he may have), he would still have needed sufficient time to change course because of the enormous momentum the ship had built up. In a similar context, global energy, economic, and ecological problems that have been accumulating gradually over many years (if not decades) are not going to be responsive to sudden changes. Moreover, there is a point-of-no-return beyond which changing course is simply not possible. Once that threshold is passed, the force of inertia will result in an uncontrollable disaster.

## Conclusions

Perhaps the most important lesson of exponential growth is that it is later than we think. In the next chapter, the impact of exponential growth on conventional fossil fuel and nuclear resources will be discussed in more detail. It is important to keep in mind the fact that making fundamental changes in something as basic as the global energy infrastructure is going to take decades, and as the fossil fuel and uranium reserves are consumed, it is only reasonable to expect their prices to increase as the global environmental problems continue to worsen.

Such is the nature of the problems that now confront the global human community.

We are not talking about minor considerations. We are talking about a systems collapse on a scale so massive it is difficult, if not impossible, to comprehend. But comprehend we must if we and our children are to be survivors. The most important consideration is that there are viable alternatives to fossil fuel and nuclear energy systems, and that the primary obstacles to developing such renewable systems are not technical, or even economic. In the final analysis, they are educational and political.

This Tenneco drilling rig made a significant discovery of oil and natural gas on Sabine Pass Block 13 off the Texas-Louisiana coast in the Gulf of Mexico.

Figure 3.1: Offshore Drilling:
Finding the last of the last of the oil.

Chapter 3

# CONVENTIONAL ENERGY CONSIDERATIONS

*"Basically the problem of providing adequate resources for the perpetuation and expansion of civilization is the problem of providing adequate quantities of energy of the right type at the right place at the right time.   This is true no matter whether the resource is food, minerals, structural materials, clean air, or energy itself."*

Harrison Brown
*Energy in Our Future* [1]

*Interrelationships*

It is significant to realize that many of the global environmental and economic problems are to a large extent related to what types of energy are used and how that energy is generated, transported and stored.  Some of the obvious environmental concerns discussed in Chapter 1 include ground, air and water pollution -- which includes acid deposition, carbon dioxide and tropospheric ozone accumulations, oil spills, and gasoline seepage from storage tanks.  Other significant environmental concerns include strip-mining, preservation of the remaining wilderness areas and the production of radioactive wastes.  Economic considerations include unemployment, budget deficits, inflation, interest rates, foreign trade deficits, and loans to developing Third World countries.

These interrelationships should not be surprising given the fact that environmental, economic, political, military, social, and industrial systems are all dependent upon stable energy resources. As such, they are secondary systems that are all predicated on a primary energy system. It follows then, that if the basic energy system is in trouble, all of the secondary systems will be also. To concentrate only on the secondary systems is like treating the symptoms of a disease rather than the cause.

Dr. Edward H. Thorndike, professor of physics at the University of Rochester, has made the observation that although increasingly scientists and engineers are being called upon to give advice on energy questions, few of them have adequate knowledge that extends beyond their own areas of specialization. This is compounded by the fact that environmental and energy problems have extensive interconnections, which means a broad knowledge of the "big picture" is at least as important as a detailed knowledge of any one aspect of it[2].

This is no small consideration. There is a general principal in education whereby as one advances, one's area of specialization invariably narrows. This results in a great many highly-educated specialists who know a great deal about their particular area of interest, but little or nothing about the infinite number of other subjects that they simply don't have time to investigate. The comprehensive analysis and communication of data can be a critically important function, yet few professionals are able to follow the events occurring outside their immediate work. This communications problem is a principal obstacle that has thus far prevented the organization and implementation of a global reindustrialization effort around renewable energy resources.

People who think the energy crisis is over do not realize that everything they now buy has been impacted by the dramatic increases in oil prices that occurred in 1973 with the Arab oil embargo. Prior to the embargo, oil was selling for about three dollars per barrel. After the embargo, the average price of oil quadrupled to about twelve dollars per barrel. Although oil prices have fluctuated dramatically since 1973, rising to in excess of thirty dollars a barrel by 1980, oil prices have never returned to their pre-embargo levels, in spite of a worldwide oil surplus. If a person buys a new automobile, the average price is now about $10,000 to $20,000, whereas the same basic vehicle could have been purchased prior to the 1973 Arab oil embargo for about one-quarter that amount. Another obvious example is the cost of housing. Homes that used to sell for $25,000 before the oil price hikes now cost around $100,000. Numerous other examples listed in Figure 3.2 illustrate how prices quadrupled in correlation with the price of oil.

Increases In Costs of Goods in Past 20 Years.

| Examples | Average Costs In: 1967 | 1987 | Percent Increase |
|---|---|---|---|
| Mo. Housing Exp. | $115.00 | $675.00 | 580% |
| Mo. Car Exp. | 85.00 | 350.00 | 410% |
| Loaf of Bread | .25 | 1.00 | 400% |
| Coffee (2 lbs) | .50 | 2.50 | 500% |
| Candy Bar | .10 | .35 | 350% |
| Man's Dress Shirt | 5.00 | 28.00 | 560% |
| Gasoline (per gal.) | .23 | 1.20 | 520% |
| Mo. Electric Bill | 128.00 | 400.00 | 312% |
| College Tuition | 300.00 | 1,580.00 | 520% |
| Postage | .05 | .22 | 440% |

Figure 3.2: Inflation.

Figure 3.2 provides examples of how the purchasing power of the U.S. dollar has fallen dramatically in the twenty-year period from 1967 to 1987. While the average cost of living increased by more than 400 percent, in many cases personal income actually declined. This "downward mobility" has increasingly become a way of life in the U.S. and much of the rest of the world.

There are endless examples of the relationship of energy costs to product costs because there is no product produced that does not have a fundamental energy cost factor. Moreover, when the cost of energy increases, inflation increases, and since interest rates must always be several points higher than the rate of inflation, interest rates go up as well. This explains why both inflation and interest rates have come down with the price of oil since 1980, in spite of record Federal deficits. But as Figure 3.3 illustrates, the current oil glut is temporary.

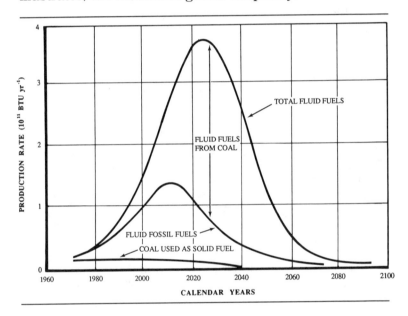

Figure 3.3: Projected Production of the World Fossil Fuels. (Prepared by Department of Physics, Texas Tech University, Lubbock, Texas[3])

At present, there are few citizens in the U.S. who realize that at current rates of consumption, the bulk of existing U.S. oil reserves will be virtually exhausted in roughly 10 to 15 years. Although new oil reserves will certainly be found, it is highly unlikely that new discoveries will keep pace with the increasing demand. As a result, it is only a question of time before oil prices begin to increase again, and with higher energy costs, inflation and interest rates will surely increase as well. The cost of energy, like any other product in the economy, is sensitive to the forces of supply and demand. As a result, oil prices can be expected to fluctuate in the short term. Because oil and other fossil fuels are nonrenewable, however, it is reasonable to assume that while their prices will vary with market forces, their overall costs will continue to increase as their reserves are depleted. This, in turn, means that the longer a transition to renewable resources is delayed, the more expensive the transition will be.

*Energy Basics*

According to data provided by the U.S. Energy Information Administration (Washington, D.C.), the principal sources of the world's energy in 1986 were liquid fuels (crude oil and natural gas liquids), natural gas, coal, and electricity from hydropower and nuclear power. Measured in British thermal units (Btu), the total world production of energy exceeded 314 quadrillion (i.e., quads) Btu[4]. (*Note*: a quadrillion is a 1 followed by 15 zeros). Crude oil and natural gas liquids accounted for 126 quads, which is more than 40 percent of the total. Coal production of 88 quads accounted for 28 percent of the total. Dry natural gas accounted for about 63 quads, or 20 percent of the total. Hydroelectric power accounted for 21 quads, which accounts for about 7 percent of the total, and nuclear power contributed 16 quads, or about 5 percent.

The leading energy producer in 1986 was the Soviet Union, which produced about 66 quads of energy. The U.S. was second with 64 quads. Together the two superpowers accounted for about 42 percent of all energy produced. The third ranking energy producer was China, which contributed about 8 percent of the world's output, followed by Saudi Arabia and Canada, which each contributed about 4 percent, and the United Kingdom which contributed about 3 percent.

All current and potential sources of energy can be classified as either renewable or nonrenewable. Nonrenewable sources include the fossil fuels, such as oil, coal and natural gas; while renewable sources generally refer directly or indirectly to solar energy. Indirect solar resources include hydropower (dams), wind, wave or ocean thermal energy conversion (OTEC) systems. Direct solar energy systems include photovoltaic cells, line and point-focus (dish) concentrators, and flat-plate collector systems. While it can be argued that the fossil fuels were also initially generated by solar energy, in that they are the products of green plants, the process of biological decay which formed the fossil fuels took millions of years. From a practical standpoint, therefore, they must be considered as a nonrenewable resource.

A key point to remember in energy discussions is that although electricity accounts for about 30 percent of most industrial countries' energy consumption, it only provides about 10 percent of the net energy use. This is because about two-thirds of the energy generated at electric utility power plants is lost to waste heat and line transmission losses. In contrast, fossil-based combustion fuels such as gasoline, coal, or natural gas account for about 90 percent of U.S. energy consumption. Even the bulk of electricity (about 70 percent) is generated from the burning of fossil fuels. In the U.S., coal provides about 57 percent of the energy consumed by electric utilities, natural gas provides about 10 percent and petroleum provides about 4 percent. Nuclear fission power plants generate about 20 percent of the electricity generated and hydroelectric dams provide about 10 percent.

The above consideration is important because in the U.S., coal is the most abundant of the fossil fuels, and its reserves have been estimated by many "experts" to be sufficient to provide for the U.S. energy needs for roughly 500 years. However, this supposed long-term availability is based on the assumption that coal is used almost solely for electricity generation (as it is today). This, of course, implies that there will be some other vast source of energy that will be used to power automobiles, aircraft, and industries that use combustion fuels, and not electricity. But even it were technically possible to mine and utilize enough coal to replace oil and natural gas, assuming an annual growth rate of 7 percent, *the 500-year supply of coal would then be reduced to about 60 to 70 years,*

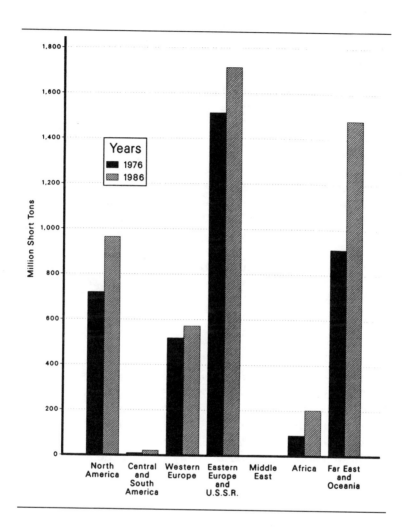

Figure 3.4: World Coal Production.
Comparing the years 1976 and 1986.
Source: U.S. Energy Information Administration[5].

In addition, according to Dr. Donald F. Othmer, an engineering professor at the Polytechnic Institute of New York (Brooklyn), a reindustrialization effort to convert coal, shale oil and tar sands into useable liquid fuels is a massive industrial undertaking. He estimated that the engineering effort alone would require virtually all of the 25 or 30 thousand competent engineers in the U.S. working for 40 years and the overall cost would easily exceed a trillion dollars[6]. In addition, the environmental impact of a major transition to coal (in terms of strip mining and carbon dioxide emissions) would be devastating.

*Crude Oil Reserves*

World crude oil reserves were estimated to be nearly 700 billion barrels as of January 1, 1987. More than half (57 percent) were found in the Middle East. The countries with the largest reserve totals, in order were:

1. Saudi Arabia
2. Kuwait
3. Soviet Union
4. Mexico
5. Iran
6. Iraq

These six countries account for 68 percent of the world's crude oil reserves. The U.S. now ranks eighth in world crude oil reserves, accounting for about 4 percent of the world total with about 28 billion barrels. Since U.S. consumption is about 2.8 billion barrels of oil annually, simple division indicates that if the U.S. did not import foreign oil, its own reserves of oil will be exhausted in about 10 years.

$$\frac{28 \text{ billion barrels}}{2.8 \text{ billion barrels}} = 10 \text{ years}$$

A great many numbers are used in discussions of energy reserves, and there is always the question of whose numbers are correct. In the past, it was a common practice for energy and natural resource analysts to tabulate all of the known oil reserves and divide the sum by the current rate of consumption to determine how long the re-

source would last.   Using this method has resulted in
many inaccurate forecasts.

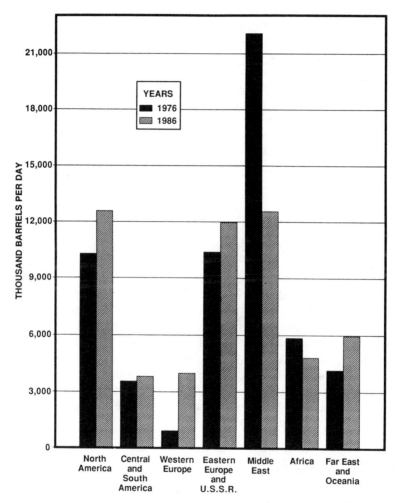

Figure 3.5: World Crude Oil Production, 1976 and 1986.
Source: U.S. Energy Information Administration [7].

For example, predictions were made in the 1920's that
the world's known oil reserves would be exhausted in 15 to
20 years (i.e., by 1940).  The world did not run out of oil in
1940 because new discoveries of oil increased the known

reserve base. It is repeatedly assumed, based on experiences such as this, that as long as the free market is allowed to operate, new discoveries of oil will be able to continually offset the increasing levels of consumption.

One of the obvious problems with this estimation method is that the reserves used in the analysis were only those that were actually discovered. In contrast, when Dr. M. King Hubbert, a recognized world authority on the estimation of energy resources undertook his analysis of the world's fossil fuel reserves, he carefully estimated the *ultimate* total production of the fossil fuels (i.e., the total size of the resource that is recoverable). This "ultimate total production" of the resource will not essentially change, regardless of when or how many of the reserves are actually discovered. In addition, the ultimate total production will not be affected by free market forces or government action or inaction.

It is for these reasons that Dr. Albert Bartlett and his colleagues at the department of physics at the University of Colorado have used Hubbert's data in calculating how long the U.S. and global fossil fuel reserves might be expected to last. Bartlett has been carefully analyzing energy and natural resource data for many years, and utilizing information from Hubbert and others, he has been able to calculate how long the U.S. and world fossil fuel reserves will be expected to last, given various rates of growth in consumption of the resources. One of the most significant aspects of Bartlett's calculations is what he refers to as as the EET of a given resource[8].

### Exponential Expiration Time (EET)

Bartlett emphasizes that growth is used as the primary indicator of economic progress. The growth of production, consumption and the Gross National Product (GNP) is the central theme of the U.S. economy, and it is regarded as disastrous when actual rates fall below projected levels. Because growth has become synonymous with success, it is important to calculate the life expectancy of energy resources given various assumptions about growth. Of special significance is the period of time that is necessary to consume a given resource, which Bartlett refers to as the "exponential expiration time" (EET) of the resource.

Table 3.1: Exponential Expiration Times (EET)
for Various Rates of Growth. (Reprinted with Permission
from Dr. Albert Bartlett)

| Annual Growth Rate | Lifetimes in Years | | | | | | |
|---|---|---|---|---|---|---|---|
| 0% | 10 | 30 | 100 | 300 | 1,000 | 3,000 | 10,000 |
| 1% | 9.5 | 26 | 69 | 139 | 240 | 343 | 462 |
| 2% | 9.1 | 24 | 55 | 97 | 152 | 206 | 265 |
| 3% | 8.7 | 21 | 46 | 77 | 115 | 150 | 190 |
| 4% | 8.4 | 20 | 40 | 64 | 93 | 120 | 150 |
| 5% | 8.1 | 18 | 36 | 56 | 79 | 100 | 124 |
| 6% | 7.8 | 17 | 32 | 49 | 69 | 87 | 107 |
| 7% | 7.6 | 16 | 30 | 44 | 61 | 77 | 94 |
| 8% | 7.3 | 15 | 28 | 40 | 55 | 69 | 84 |
| 9% | 7.1 | 15 | 26 | 37 | 50 | 62 | 76 |
| 10% | 6.9 | 14 | 24 | 34 | 46 | 57 | 69 |

The EET is calculated by knowing the size of the resource, the rate of its use, and the fractional growth-per-unit time of the rate of consumption. While most resource scholars may be familiar with this equation, Bartlett has observed that there is little evidence that it is known or understood by the political, industrial, business, or labor leaders who continue to emphasize how essential it is for our society to have uninterrupted growth.

Table 3.1, prepared by Bartlett, reveals how important the relationship is between the lifetime of a resource and the rate of consumption. Note, for example that a resource that will last for 1,000 years at current rates of consumption (zero growth) will be exhausted in 115 years if the annual growth increases by only 3 percent. According to data compiled by Hubbert, world oil production from

1890 to 1970 has grown at a rate of about 7 percent, which results in a doubling time of 9.8 years. Given a historic growth rate of 7 percent, consider Tables 3.2 and 3.3, also prepared by Bartlett:

Table 3.2:
Exponential Expiration Time (EET)
of U.S. Oil Reserves.

| Col. 1 (% Growth) | Col. 2 (yr.) | Col. 3 (yr.) | Col. 4 (yr.) |
|---|---|---|---|
| 0% | 2006 | 2009 | 2041 |
| 1% | 2003 | 2005 | 2027 |
| 2% | 2000 | 2002 | 2019 |
| 3% | 1998 | 2000 | 2013 |
| 4% | 1997 | 1998 | 2009 |
| 5% | 1996 | 1997 | 2006 |
| 6% | 1995 | 1996 | 2004 |
| 7% | 1994 | 1995 | 2002 |
| 8% | 1993 | 1994 | 2000 |
| 9% | 1992 | 1993 | 1999 |
| 10% | 1991 | 1992 | 1998 |

Column 1 in Table 3.2 is the percent annual growth rate. Column 2 represents the year that the estimated oil left in the lower 48 states will be essentially depleted. Column 3 includes Alaskan oil reserves (about 10 billion barrels) along with the reserves of the lower 48 states for a total of 103.4 billion barrels. Column 4 includes the estimates of U.S. oil shale (103.4 billion barrels) in addition to the oil contained in Alaska and the lower 48 states, for a grand total of 206.8 billion barrels.

Table 3.3 provides a summary of the estimates of world oil reserves. Column 1 is the percent annual growth rate of world oil production. Column 2 provides the approximate year that the estimated crude oil reserves left in the world as of 1978 (estimated to be 1,691 billion barrels) will be effectively gone. Column 3 includes the potential oil shale (approximately 190 billion barrels) plus the remaining crude for a total of 1,881 billion barrels. Column 4 assumes that the amount of shale oil is four times the amount that is now known, thus bringing the total to 2,451 billion barrels.

Table 3.3:
Exponential Expiration Times (EET)
of World Oil Reserves.

| Col. 1 (% Growth) | Col. 2 (yr.) | Col. 3 (yr.) | Col. 4 (yr.) |
|---|---|---|---|
| 0% | 2079 | 2091 | 2125 |
| 1% | 2048 | 2053 | 2068 |
| 2% | 2033 | 2037 | 2047 |
| 3% | 2025 | 2027 | 2034 |
| 4% | 2019 | 2021 | 2026 |
| 5% | 2014 | 2016 | 2020 |
| 6% | 2011 | 2012 | 2016 |
| 7% | 2008 | 2009 | 2013 |
| 8% | 2006 | 2007 | 2010 |
| 9% | 2004 | 2005 | 2008 |
| 10% | 2002 | 2003 | 2004 |

When Bartlett initially published Tables 3.2 and 3.3, he used periods of time (i.e., the number of years from 1978) and not the specific dates listed. This is because by listing a specific date, one assumes the variables, including the rates of growth, will remain constant, which is highly unlikely. There is also the fact that oil and other fossil fuels will never be totally exhausted. It is just a question of how soon they become too costly to extract and use what is left. For whatever reason, if the estimated date passes and there is still oil available, people will undoubtedly say "*See -- he was wrong.*" Nevertheless, by listing the actual years, one is able to compensate for the fact that the initial estimates were made more than 10 years ago, and thus far, Bartlett's projections of when U.S.

and world oil reserves would be expected to be depleted have, if anything, turned out to be optimistic.

For example, according to an estimate that was issued by researchers at the U.S. Geological Survey (USGS), in September of 1985, world oil reserves totaled only 723 billion barrels (i.e., about a 36-year supply). The report estimated undiscovered resources at 550 billion barrels, for a total of 1,273 billion barrels[9]. This is in contrast to the 1,691 billion barrels that Bartlett assumed was available in 1978. Although about 17 billion barrels have been consumed annually for the 7 year period since 1978 (which accounts for some 120 billion barrels), the USGS report still assumed world oil reserves to be about 300 billion barrels *less* than Bartlett's assumptions.

Other investigators have reached similar conclusions with respect to fossil fuel reserves. Two studies in particular are worthy of mention. The first was a two-year international effort under the direction of Dr. Carroll Wilson of the Massachusetts Institute of Technology (MIT). Some 70 energy analysts were recruited from government and the universities in fifteen major noncommunist countries to focus on calculating energy supply and demand. Their conclusions of the studies were as follows:

1. There is only a finite amount of oil and there are limits to the rate at which it can be recovered. Sometime before the year 2000, the decreasing supply of oil will fail to meet the increasing demand.

2. Because large investments and long lead times are required for developing new energy resources, it is important that the effort begin immediately[10].

The second two-year-long study was undertaken at the request of the USGS and the Department of Energy by the RAND Corporation, a highly respected research group based in Santa Monica, California. The conclusions of the 700-page report that was published in 1981, were that the prospects of finding more oil and gas in the U.S. are severely limited. Moreover, the reason is geology, and not economics. The U.S. is simply running out of unexplored places where there is any possibility of finding significant amounts of oil[11].

Tables 3.1, 3.2 and 3.3 clearly demonstrate that *even a doubling of the resource only results in a small increase in its life expectancy given the nature of exponential growth.* This hard reality is not overlooked by senior executives of the major oil companies. In an essay published in *Newsweek* magazine, Allen E. Murray, president of the Mobil Oil Corporation, repeatedly warned against complacency. He indicated that new oil shortages were inevitable, the only question was when?[12] Other experts outside of the oil industry have made similar warnings about declining U.S. oil reserves. As the data in Table 3.3 indicate, calculations of world oil reserves are not encouraging, particularly in the face of growing levels of consumption by Third World countries who are only beginning to industrialize their societies.

In sharp contrast to these perspectives are the beliefs of many people who assume that there are enough fossil fuel reserves to last for thousands of years. As Bartlett indicates, however, it is possible to calculate an *absolute upper limit* to the amount of oil the Earth could contain. He simply asserts that the volume of oil in the Earth cannot be larger than the volume of the Earth itself. The volume of the Earth is roughly equivalent to about (6.81 x $10^{21}$) barrels. As a result, if the rate of oil consumption in 1980 were to continue with an annual increase of about 7 percent, an entire Earth full of oil would be consumed in only 342 years.

## The Oil Surplus

Geophysicists were predicting as far back as World War II that by the year 2000, U.S. oil reserves would be unable to sustain projected consumption requirements. But with substantial discoveries in the Middle East and elsewhere, the U.S. has been able to import (at great cost) sufficient quantities to offset declining domestic production. U.S. oil production peaked in 1970 and has been declining ever since. The Soviet Union is currently the world's largest oil producer, but Soviet oil production peaked in 1984 and has also been in the process of declining. Thus in the future, both superpowers will be increasingly dependent on the remaining oil reserves in the Middle East countries.

When the Middle East Arab oil embargo occured in 1973, the price of oil increased by a factor of four (from $3 per barrel to about $12 per barrel). In addition to the psychological shock of not being able to obtain gasoline without great difficulty, a profound economic transition was set in motion by the oil price increases. Automobiles and houses began to get smaller and much more expensive, but something much more ominous began to happen to many of the energy-intensive steel and other smokestack industries, which had been the backbone of American industry. They went out of business.

Prior to 1973, the majority of U.S. oil consumption was used by heavy industry. As a result, when energy prices quadrupled, many of these industries were unable to remain competitive, and were forced to shut their doors. In the world-wide recession that followed, millions of workers were forced on the unemployment rolls as industry after industry shut down its U.S. operations. Nor was it just industrial workers who were affected. Many third-generation farmers and fisherman have also been forced out of business because the cost of running their diesel-powered machines exceeded the income generated by their harvests. It was, however, the loss of heavy industry, along with a world-wide recession, increases in Arab oil production, and the development of more energy-efficient automobiles, homes and appliances, which collectively resulted in the oil surplus which began in 1980.

One obvious impact of the global oil surplus was that the price of oil began to decline (although it has never fallen below the twelve dollar per barrel levels that were established after the initial Arab oil embargo). This, in turn, explains why double digit inflation rates began declining to around 4 percent, which in turn explains why the economy began to revive in 1982. In spite of these real world physical factors, much of the U.S. media and the general public assumed that it was President Reagan's policies of cutting taxes, while dramatically increasing annual defense expenditures, that were primarily responsible for the economic upturn.

While it is true that the substantial increases in defense spending contributed to stimulating the economy, the net result was to create the largest deficits in U.S. history. This has been compared to living off of one's credit card. It is a short-term band aid solution that will only allow the fundamental problems to worsen. Such stopgap measures will work for awhile, but eventually the bill comes due, and when that happens, the house of cards falls apart. The bottom line on the current surplus of oil is that it is temporary. The initial gas lines during the 1970's were merely a warning sign that there is much more serious trouble ahead. Fortunately, the current global oil surplus has provided some breathing space to organize an energy transition without the panic reactions that usually result from waiting until the situation goes critical.

It is for this reason that it is especially unfortunate that senior officials in both the Reagan and Bush Administrations have apparently not understood that humanity is confronted with unprecedentedly serious energy and environmental problems. Rather, both Administrations have generally held the view that environmental concerns must give way to commercial interests, and that sufficient oil and coal will be found in the remaining wilderness and offshore areas in the U.S. and elsewhere. As a result, instead of using the limited time available to make an orderly transition to renewable energy resources, a decision was made to essentially dismantle the existing renewable energy research programs and exploit what is left of the American offshore and wilderness areas. This, in turn, has only accelerated the exponential destruction of the remaining wilderness and wildlife habitats.

*The Battery*

The fossil fuels are like the battery in an automobile. There is enough stored energy in the battery to start the engine, but if one tries to propel the vehicle on the stored energy in the battery alone, one is not going to get very far. The vertical numbers in Figure 3.6 represent trillions of kilowatt hours per year. The horizontal numbers are in thousands of years, from minus 5000 to plus 5000 years from the present. Perhaps the most significant consideration of Figure 3.6 is that the petroleum era will be a temporary one, yet the global environmental damage it is causing could be irreversible. The survival of the human species lies in the balance.

Figure 3.6: The Battery.
The Petroleum Era in a Historical Perspective.
(Illustration prepared by Hubbert[13])

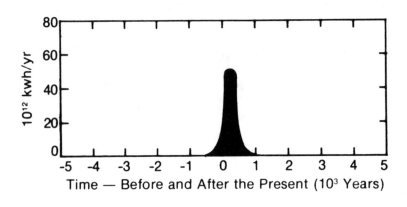

Time — Before and After the Present ($10^3$ Years)

There is also the hard reality that if there is a continued insistence on using the remaining fossil fuel reserves to provide fuel for industrial and vehicular engines, after a certain point, the fossil fuels will be quickly depleted. At that point, little -- if any -- economically recoverable energy reserves will be left to build the renewable resource infrastructure. That, in turn, means the cost of the transition will be many times greater.

There are many individuals who are either unaware of, or are unconvinced of the hard realities of exponential growth. They are probably not aware of the studies done by the U.S. Geological Survey, M.I.T., or the RAND Corporation, that clearly warn of the dangers of just continuing to consume the remaining fossil fuel reserves. They are also apparently not aware of the many studies that have documented the serious global environmental hazards that are associated with using fossil fuels for energy generation purposes. To ignore such significant scientific information is analogous to the captain of the *Titanic* refusing to acknowledge the warnings about the icebergs.


## Nuclear Power

Nuclear-fission power plants were initially thought to be the answer to the diminishing fossil fuels. Even though the enriched uranium fuel was also severely limited, it was assumed that a more advanced nuclear technology -- referred to as "breeders" -- would eventually be made commercially viable. Breeder reactors would actually be able to produce more radioactive "fuel," in the form of plutonium, than they consume. As a result, their plutonium fuel would be renewable. This was a great concept in engineering theory, although biologists had continually warned that plutonium is one of the most toxic elements known, it is very difficult to handle, and it would remain deadly for about 250,000 years. In addition, if the "plutonium economy" was ever to become a reality, millions of tons of plutonium would have to be produced and shipped throughout the U.S.

In spite of these long-term and exceedingly difficult and dangerous environmental concerns, conventional nuclear reactors and their breeder offspring constituted America's primary energy strategy since the 1950's to resolve the diminishing fossil fuel problem. However, when the partial meltdown accident occurred in 1979 at one of the new nuclear reactors at Three Mile Island in Pennsylvania, both public and investor confidence in nuclear fission technologies were shattered. And although billions of taxpayers' dollars have been used to develop and promote nuclear energy systems, rather than solving the diminishing fossil fuel problem, nuclear technology has instead created an even more awesome problem of its own -- radioactive waste.

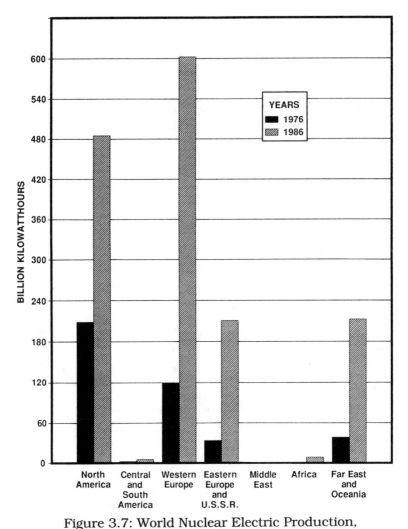

Figure 3.7: World Nuclear Electric Production,
1976 and 1986.
Source: U.S. Energy Information Administration [14].

The radioactive waste problem is unique for many reasons, but one of its most insidious aspects involves the fact that it is invisible to the human senses until disease or death occurs. When moisture is present, radioactive isotopes spread in an ecosystem like red dye spreads in a glass of water, and the isotopes will remain toxic for anywhere from a few days to in excess of a million years, depending on the isotope. There is still no long-term storage plan for these waste products and many of the existing waste storage facilities are full and out of control in terms of their ability to prevent the radioactive wastes from leaking and spreading.

This "spreading" problem occurs because under irradiation, all materials change their nature as their atoms become unstable. The world of radioactivity is a world of continuous change. This is why the storage vessels which contain the radioactive wastes are only reliable for short periods of time. Eventually, they too will become radioactive. This, in turn, means the longer a nuclear reactor is operating, the more radioactive it becomes. This is the primary reason why any repair or maintenance of aging nuclear reactors is an extraordinarily hazardous task.


*Radioactivity*

The problem of radioactivity in the reactor and the surrounding building is in addition to the liquid, gaseous and solid waste generated by the uranium fuel rods. As the uranium undergoes fission, the uranium atoms split, and in so doing, release neutrons. Some of these neutrons split other uranium atoms, which, in turn, produce the radioactive waste products. The net result of this fission process is the generation of the intense heat that is used to generate steam for the production of electricity.

The difference between a nuclear reactor and a nuclear weapon is measured by the number of neutrons that are released in the fission process over a given period of time. If only a limited number of neutrons are available for triggering the fission chain-reaction, the reaction can be controlled for energy production purposes. If too many neutrons are released, the chain-reaction will rapidly accelerate, resulting in an atomic explosion. To prevent this from happening, nuclear reactors have control rods and water circulation to regulate the fission process by absorbing the extra neutrons.

However, during normal operations, some of the neutrons that are released pass beyond the uranium and into the steel structures which hold both the fuel assemblies and the cooling water which flows between them. Other neutrons penetrate the massive concrete shielding outside the steel reactor vessel. These neutrons are absorbed by the atoms of iron, nickel and other elements that make up the steel, water and concrete. When atoms absorb neutrons, they are rendered "unstable" (i.e., radioactive) for various lengths of time. In the case of nickel-59, which has a half-life of 80,000 years, it will need to be shielded from humans for about a million years[15].

*Decommissioning*

Decommissioning is the term used to describe what happens to a nuclear reactor complex once its theoretical useful life of 30 or 40 years has been completed. This "back end" of the nuclear fuel cycle is generally not discussed because no one really knows how such a difficult task is to be accomplished, or what it will ultimately cost over hundreds of thousands of years. Utilities and waste processing companies are usually not concerned about the long-term waste storage costs because in most cases, they have no long-term legal or financial responsibility to manage the radioactive wastes. That responsibility is given to the U.S. taxpayers of the future. But if the experience of attempting to decommission a nuclear fuel recycling plant at West Valley, New York is any indication of what is ahead, there are plenty of reasons to be concerned.

The $32.5 million West Valley plant, located 30 miles southeast of Buffalo, was officially opened with much fanfare in June of 1963, although it did not actually begin to reprocess nuclear wastes until 1966. But it was only to operate for 6 years before its operator, Nuclear Fuel Services (NFS), a subsidiary of W.R. Grace's Davison Chemical Company, abandoned the facility. Left behind were 2 million cubic feet of buried radioactive trash and 600,000 gallons of highly radioactive liquid waste that is now seeping into the Cattaraugus Creek, which flows into Lake Erie, from which the city of Buffalo obtains its drinking water. The cost of cleanup is estimated to be at least $1 billion, assuming such a cleanup is even possible.

The West Valley plant was the world's first commercial nuclear-waste facility that would be able to take the spent fuel from nuclear power plants and reprocess it for renewed use. This recovery and reprocessing of spent fuel rods is an important step in the nuclear fuel cycle because only about 1 to 2 percent of the nuclear fuel is initially consumed in most commercial reactors, and uranium is not a renewable resource. This is particularly true of the reactors in the U.S. that require an energy intensive, highly enriched uranium 239. Unfortunately, the problems associated with trying to handle highly radioactive wastes proved to be overwhelming to the workers and management at the West Valley facility. An extensive investigative report on the West Valley plant titled "Too Hot to Handle," undertaken by *The New York Times*, concluded the following:

> *"It is the story of technocrats who assured and reassured the public that nuclear recycling was safe and that a thoughtfully engineered fail-safe system would minimize the hazards of any accidents that might possibly occur -- without making it clear that their assurances were based on extrapolations from premises rooted in probabilities and anchored in uncertainty. It is the story of company officials who repeated such assurances even after scores of incidents -- known only inside the company and to a few governmental inspectors -- had made it clear that leakage of radioactivity in the plant was reaching dangerous levels."* [16]

Radiation at the West Valley facility was rapidly spreading, and the costs of operation were dramatically increasing. Finally, in 1975, NFS announced that it would have to spend at least $600 million to make the facility manageable, which was nearly twenty times the initial capital cost of the plant. This, plus the fact that operating costs had increased by some *4,300 percent* since the plant began operation, caused NFS to shut the facility down, and let the taxpayers of New York figure out what to do with the multi-billion dollar waste storage problem that was left behind [17].

*Nuclear Waste Storage*

As it turns out, the West Valley facility in New York is merely the tip of the nuclear waste iceberg. The U.S. Department of Energy (DOE) spent years and $700 million to build the nation's first permanent waste storage facility deep in salt beds near Carlsbad, New Mexico. But only after the facility was completed in 1988 did the DOE officials discover that water had somehow leaked into the salty caverns. As a result, the Carlsbad facility has been put on hold, but the nuclear waste problem in the U.S. is so critical that Idaho Governor Cecil Andrus "declared war on the Department of Energy" in October of 1988. Under the governor's orders, state police halted any further shipments of nuclear waste into the State of Idaho [18].

The nuclear waste problem has been around for decades, and in spite of the fact that thousands of brilliant scientists have been working on the problem, no one in any country has come up with an acceptable long-term solution. Moreover, as the general public learns more about the waste problems, they become more and more convinced that they don't want the toxic dump in their state. Given this bleak outlook, the U.S. Congress decided just hours before the Christmas adjournment in 1987 that a place called Yucca Mountain, located in a remote nuclear test site in Nevada, was to be the nation's final resting place for high-level radioactive waste. Not only was the area already radioactive, but Nevada simply had less clout in Congress than any other state. However, there are two serious environmental problems associated with the Yucca Mountain site.

First, scientists point out that the Yucca Mountain area is in a geologically unstable region with active volcanos and earthquakes. Second, and perhaps more worrisome, is the fact that the containers for the nuclear waste are only expected to last for 300 to 1,000 years. After that, it must be the mountain itself that contains the waste products, and therein lies the problem. Scientists from the U.S. Geological Survey had drilled two small shafts in Yucca Mountain and they found that the mountain "breathes." In the winter, warm air flows out of the shafts from the center of the mountain. In the summer, the pattern is reversed and the air is sucked into the mountain. In a memorandum, 17 scientists from the USGS stated that this air flow could release radionuclides into the atmosphere.  In spite of the scientific warnings, Yucca Mountain is still viewed the primary designated U.S. waste storage facility[19].

If that were not bad enough, in 1989, officials from the U.S. Department of Energy announced that in spite of the critical nature of the nuclear weapons program, all three of the nuclear weapons facilities had to be shutdown due to massive technical problems associated with dealing with the radioactive waste.  Department of Energy officials have now admitted that extensive releases of radiation had often contaminated workers and large numbers of U.S. civilians who happened to be living within the proximity of these facilities.  The general public was deliberately not informed of these radiation leaks for decades because of "national security" considerations.  As one Federal official put it, the emphasis was not on safety, but on production.

It has been estimated that it will take anywhere from 100 to 150 billion dollars to clean up the existing nuclear weapons facilities, assuming such a clean-up is even possible.  In reality, such clean-ups have meant collecting and moving as much of the wastes as possible to some other location, which ultimately only creates another problem that will eventually have to be dealt with.

True clean-ups have up till now proven to be beyond the limits of existing technical and engineering capabilities.  This unresolved technical problem has led to the concept of "National Sacrifice Zones," which implies sealing-off the contaminated areas from the members of the general public, presumably for hundreds of centuries. Such a policy in and of itself, poses some interesting social and ethical problems.  For example, imagine trying to figure out how to make a sign or symbol that could explain the radioactive danger which could still be understood by our descendants thousands of years into the future.  One can only wonder what the future generations will think of this generation for leaving them such awesome problems, so that we could power items like electric can openers, neon signs and hair dryers.

*Questions of Safety*

While nuclear proponents like to point out that it is highly unlikely for a nuclear fission power plant to explode like a nuclear bomb, which of course is true, they never bother to explain that given the amount of radiation in a modern 1,000 megawatt (mW) reactor, a serious loss of coolant accident could result in a melt-down condition which could be as dangerous as an outright nuclear explosion. This is because the amount of radiation in a 1,000 mW reactor is more than a thousand times that which was produced by the atomic bomb that was dropped on Hiroshima in 1945. In a melt-down scenario, this massive amount of deadly radiation could all be released at ground-level, whereas a nuclear or conventional chemical or steam explosion would disperse much or most of the radioactive particles high into the Earth's atmosphere. This is essentially what happened when the Chernobyl accident occurred in the Soviet Union on April 26, 1986.

Although the Soviet government initially assured those living in the area of the Chernobyl accident that the medical impact would not be significant, three years after the accident cancer rates have doubled among residents of contaminated farm regions, and calves and other farm animals have been born without heads, limbs, eyes or ribs. Moscow News has reported that more than half of the children in the Narodiehsky region of the Ukraine have illnesses of the thyroid gland, which exposure to radiation can cause. Soviet officials now admit that they "drastically under-estimated" the health problems caused by the reactor explosion and fire[20].

There are also other serious problems associated with nuclear fission power plants. Once a nuclear reactor becomes operational, workmen are not able to clean and maintain the critical interior components of the reactor vessel due to the intense level of radiation. This is in contrast to the normal procedure in fossil fuel plants where maintenance personnel are able to shut the plant down periodically to clean and inspect all of the critical interior components. This regular maintenance schedule is a major reason why there are so few accidents or failures in fossil fuel power plants, in contrast to nuclear fission facilities, whose unreliability has become one of their most dependable features.

### Military vs. Civilian Reactors

In contrast to the civilian nuclear reactors that have been plagued by accidents and shutdowns, the U.S. Navy has established a respectable performance record with its nuclear surface ships and submarines. There are, however, distinct differences in the size and quality of the nuclear systems used by the U.S. Navy, in contrast to the much larger reactors engineered for commercial power production. The most obvious differences involve the reactor size, design configuration, and cost per installed kilowatt (kW). The reactors manufactured for the U.S. Navy are all designed to be relatively small; they are all based on a similar design concept; and they were initially about five times more expensive to construct (in terms of cost per kW) than their civilian counterparts.

William E. Heronemus, now professor of civil engineering at the University of Massachusetts, supervised the construction of nuclear submarines for Admiral Hyman Rickover. In 1973, Heronemus testified before a nuclear licensing hearing that it would cost roughly $2,400 per kW to build one of the U.S. Navy's nuclear reactors, compared with $400 per kW for the commercial plant that was under review. (Note that at present, capital costs for commercial reactors are closer to the $2,400 figure.) Heronemus also testified that he would have refused to approve the wiring and piping for the Navy that had been accepted and incorporated in the commercial nuclear facility.

### Advanced Reactor Considerations

According to Dr. Charles E. Till, a nuclear physicist at the Argonne National Laboratory (Illinois), a new generation of nuclear fission reactor, referred to as a "Integral Fast Reactor," has been under development by the U.S. Department of Energy for several years. This new liquid sodium-cooled reactor configuration is expected to be safer, minimize corrosion and be more efficient (i.e., it should be able to use 15 to 20 percent of the uranium fuel instead of the 1 to 2 percent with current reactors), and generate less radioactive wastes than the existing generation of light-water reactors in use in the U.S. [21].

In addition, according to Jerry Griffith, an associate deputy assistant secretary for reactor systems development for the DOE, the Integral Fast Reactor is considered especially crucial to the future of nuclear power because it is the best technology for breeding plutonium. As the world's uranium reserves become scarce, plutonium will be needed as a substitute nuclear fuel. Thus, the Integral Fast Reactor is the key to a nuclear future and a "plutonium economy." However, there are three primary concerns, with the advanced Integral Fast Reactor:

1.  Nuclear physicists have had a relatively poor track record in predicting the engineering outcome of many theoretical calculations, and thus far, not even a prototype of the Integral Fast Reactor has been tested.

2.  Liquid sodium is an extremely volatile substance that will burst into flames if it comes into contact with either air or water. Two liquid sodium-cooled U.S. prototype nuclear breeder plants were totally destroyed by liquid sodium fires.

3.  Plutonium is an exceedingly difficult and dangerous material to handle. It is one of the most toxic elements known. Plutonium is *35,000* times more lethal than cyanide poison by weight. There is an equivalent of 20 million mortal doses in only 5 grams of plutonium, which is the weight of a 5-cent coin, and it will remain dangerous for hundreds of thousands of years. This long-term toxicity of plutonium, in and of itself, creates significant moral and ethical questions about producing such long-lived toxic substances that are invisible to human senses.

It is important to realize that in order to produce enough energy to displace fossil fuel resources, which now account for roughly 90 percent of the industrial world's energy supply, immense quantities of plutonium would have to be created. And given the extremely poor track-record of containing or properly disposing of the plutonium and other radioactive wastes that have already been created, the construction of the thousands of plutonium-fueled reactors that would be necessary for the "plutonium economy" to be viable would hardly seem to be an acceptable alternative. These serious problems underscore the importance of evaluating the renewable energy options that do not pose such long-term unknown environmental and economic risks.

*Nuclear Fusion*

Nuclear fusion reactors, in contrast to nuclear fission reactors, do not split uranium atoms. Rather, they are intended to fuse hydrogen atoms in a process similar to that which occurs in the Sun and other stars. Although fusion physics is a common occurrence in stars, it is well to remember that no biological organisms are able to live in such high-temperature environments. Billions of dollars have already been spent on this highest of high-tech energy technologies, which has been under development for decades by governments in the U.S., Japan, France, Germany, the Soviet Union and other European countries. However, it has been as a result of this research that at least some fusion reactor advocates are now questioning whether such high-temperature (over 100 million degrees F) fusion energy systems will ever play any role in energy production during the next 50 to 100 years.

According to a recent nuclear fusion update article by John Horgan published in *Scientific American,* the initial hopes of having small, safe high-temperature fusion reactors burning cheap, abundant fuel have all but disappeared[22]. It is now estimated that exotic and expensive fuels will be required, they will produce significant quantities of radioactive waste, and even the smallest fusion reactor would be comparable in size and complexity to the largest of today's fission reactors. Such high-temperature fusion technologies face staggering technical problems, and billions of additional dollars will be needed just to build a prototype. Even if the prototype plant works from a technical perspective, the really important question is whether nuclear fusion systems will ever be cost effective. As such, it would be irrational to predicate a nation's energy policy on such high-risk technologies.

*Cold Fusion*

On Thursday, March 23, 1989, two scientists, Dr. Stanley Pons, chairman of the department of chemistry at the University of Utah, and his colleague, Dr. Martin Fleischmann, professor of electrochemistry at the University of Southampton, England, stunned the scientific world when they held a national press conference to announce the results of their experiments. Their announcement indicated they had not only succeeded in generating a fusion reaction that actually produced more energy than the reaction consumed, but they had accomplished it at room temperature, in a simple table-top test tube apparatus in their kitchen. They did not indicate what radioactive isotopes were produced as a result of their room temperature reaction.

The two scientists had been working on what was termed "cold fusion" by the news media, for more than 5 years, and they felt that commercial reactors based on the new low-temperature fusion process could be in operation in about twenty years. However, most scientists were highly skeptical of such claims, and they were troubled that such an announcement came at a press conference rather than having a paper published through the traditional peer-review process [23]. As a result, it is impossible to know as of this writing whether or not the cold fusion process is valid.

If Pons and Fleischmann turn out to be correct in their announcement, there is certainly the possibility that such a breakthrough could have a profound impact on global energy planning. But if the cold fusion process is able to be developed into an electrical power plant, Pons suggested it would probably be used to heat water into high-temperature steam, which would be fed into conventional steam-generation equipment. If this is the case, it is still questionable whether such cold fusion technologies could compete economically with the point-focus solar technologies discussed in Chapter 5 that could be mass-produced in automobile and aerospace industries. There is also the issue of what radioactive wastes will be generated from such nuclear reactions.

*Nuclear Economics*

The true cost of nuclear power has been confused by the quasi-public nature of the research and development. U.S. taxpayers are financially responsible for the "back-end" of the nuclear fuel cycle, which includes covering any costs not met by the utility for waste disposal and decommissioning. Billions of taxpayers dollars have also been spent for the "front-end" of nuclear research and development. These costs are not included in most nuclear cost totals. They include the construction and operation of the three U.S. uranium fuel enrichment facilities, that are at present shut down due to the extensive problems with respect to radiation spreading. When all three of these enrichment facilities were operating at full capacity, their electrical requirements were actually about the same as those used by the entire country of Australia. Other excluded costs include Federal regulation, long-term waste disposal, and the numerous health costs that are associated with people being exposed to radiation.

To comprehend the nuclear issue, it is necessary to put time in perspective. If toxic wastes, which will be deadly for 200,000 or 500,000 years, are generated, is it possible for anyone to comprehend the actual environmental or economic costs? The very first civilized groups of people in the Middle East appeared only about 8,000 to 10,000 years ago. How is it, then, that one generation could, or should, assume the right to create insidious radioactive hazards that will remain deadly for hundreds of thousands of years? How is it that we have allowed ourselves to do such things that cannot be comprehended or calculated in terms of cost or human death and disease?

The nuclear and other toxic waste problems are global in nature and have clearly transcended the capitalist or communist ideologies, as both political systems have developed nuclear technologies. It is interesting to note that both the U.S. and the Soviet Union -- as well as most other countries -- have up to now shown a complete disregard for the "human rights" of future generations. It seems difficult to imagine how so many "civilized" nations could have allowed the production of such deadly and long-lived radioactive wastes to occur. Even more difficult to understand is how the citizens in the same countries can silently let the production of such long-lived toxic wastes continue. If the ovens of the prison camps in Nazi Germany are now viewed as a moral outrage, one can only wonder how future generations will view the actions of the present generation.

Never before has the human community been faced with such an awesome array of problems, and unlike most other environmental problems, acceptable solutions for the disposal of radioactive wastes are as yet unknown. This hard reality is underscored by the fact that after more than 30 years of concentrated effort by a wide-range of distinguished scientists from around the world, no one has yet demonstrated a solution to the radioactive waste problem. Indeed, a study done by the Jet Propulsion Laboratory of the California Institute of Technology for the President's Office of Science and Technology Policy concluded the following:

> "The problems of high-level nuclear waste management are so complex and have so many ramifications that no one person or group of persons can possibly have all the answers. The results of this study indicate that the U.S. program for high-level waste management has significant gaps and inconsistencies." [24]

As a result, it would seem that far from solving the diminishing fossil fuel problem, the nuclear power industry has only succeeded in creating a whole new range of technical and long-term environmental problems that will be inherited by our children and their children for thousands of generations into the future. This unfortunate reality brings to mind the observation that Albert Einstein made about nuclear technology:

> *"The splitting of the atom has changed everything save our modes of thinking, and thus we drift toward unparalleled disaster."*

Dr. Barry Commoner made a similar observation in his book, *The Closing Circle*:

> *"Our experience with nuclear power tells us that modern technology has achieved a scale and intensity that begins to match that of the global system in which we live. We cannot wield this power without deeply intruding on the delicate environmental fabric that supports us. It warns us that our capability to intrude on the environment far outstrips our knowledge of the consequences."* [25]

Placing such an ominous issue as radioactivity into an evolutionary perspective is not easy. Kenneth and David Brower have expressed it as well as anyone in an article, "Miracle Earth" that was published in *Omni* magazine. In their article, they point out that when a beta particle, (a high-speed electron emitted from the nucleus of a radioactive atom) strikes living tissue, "it rips negatively charged electrons from the tissue's atoms, leaving positively charged ions in its wake." These liberated electrons in turn ionize other atoms in a cascading effect which tears apart tens of thousands of highly-ordered biological molecules that serve as the structure of living cells. "The passage of such a particle leaves the city of the cell in ruins. Alpha, gamma, and X-rays all have this effect on biological molecules. Their entry hole is small, but their exit hole is spectacular."

In their article, Kenneth and David Brower go on to describe the delicate balance that is held between life and radiation as follows:

*"Life is adrift in a sea of radiation, as the makers of the artificial kind are apt to point out... Radiation seeps up from springs on earth and flows in rivers from space. The sun sends out a steady stream of particles freshened occasionally by solar storms. Yet on earth, life has found something like a back-water. Here life is protected from the full force of the cosmic stream by the planet's atmosphere; here life tolerates the weaker radiations from the planetary crust... But the margin within which life operates is small, and the tolerances are fine. There is no question who the enemy is, our first and oldest. If the universe is hostile to life, there is no better expression of that hostility than the radioactive particle.*

*"The Four Horseman are secondary enemies, at a less basic level of organization. The particle attacks us fundamentally by disordering the atoms of which we are made. It strikes at and scatters the miraculous principle that distinguishes us from dust."* [26]

## Conclusions

Because fossil fuels are nonrenewable and are being exponentially consumed, they will be unable to sustain an expanding global industrial economy. Even if the fossil fuels were inexhaustible, however, their unacceptable environmental impact would still dictate that alternative energy technologies and resources be developed.

Because nuclear energy systems do not produce greenhouse gases or acid deposition, they are being touted as the logical solution to the problems created by burning fossil fuels. But as this chapter has documented, nuclear fission technologies generate staggering environmental problems of their own that revolve around the radioactive wastes that have thus far proven to be unmanageable. Moreover, if nuclear power plants were to effectively replace the burning of fossil fuels, literally thousands of nuclear reactors would have to be built at a cost of trillions of dollars. In addition, since the existing uranium reserves will barely be able to keep the 110 existing reactors in the U.S. operating beyond the year 2000, a whole new type of untested breeder nuclear technology would have to be rapidly developed.  For all of these reasons, if nuclear systems were the only energy alternatives to the burning of fossil fuels, there would be little reason to be optimistic about the future of the human community.

Fortunately, there are several viable renewable solar-hydrogen technologies discussed in Chapters 4 and 5, and the renewable resources discussed in Chapter 6, that are capable of displacing the use of both fossil and nuclear fueled energy systems.  While the political aspects of re-ordering national priorities are very real, most elected officials determine their issue priorities by finding out what the majority of their constituents think is important. Although elected officials are often criticized for tailoring their views to the whims of public opinion, they do have a responsibility to be aware of how the majority of the people they represent feel about issues. This essentially means it is necessary for the majority of voting citizens to be informed about two important things:

1. The global energy, environmental, and economic problems are continuing to worsen exponentially.

2. An industrial transition to renewable solar-hydrogen resources represents a fundamental solution to many of the most serious energy and environmental problems that threaten the survival of the Earth's ecosystems.

Once the majority of the voting public understands these two key points, a bipartisan political mandate to re-order national priorities around a transition to renewable energy resources can be established.  Given the serious nature of the many environmental problems, such a transition needs to be undertaken with wartime speed.

Figure 4.1: Stellar Hydrogen.
Giant molecular-cloud complexes, consisting almost entirely of hydrogen, are the most massive objects within galaxies. Gravity is the primal force that eventually causes the hydrogen to compress until it can fuse into the heavier chemical elements. As such, hydrogen is not only the primary fuel for the Sun and other stars, it is also a primary building block of matter in the known universe.
(Courtesy of National Optical Astronomy Observatories)

Chapter 4

# Hydrogen

*In the Beginning*

Since the very beginnings of recorded history, there have been people who have revered the Sun as the source of life. This should not be surprising, because it is fairly obvious that without the energy emitted by the Sun, life as we know it could never have evolved on the Earth. Few people, however, realize that the primary fuel for the Sun and other stars is hydrogen.

Although the primal force that causes the Sun and other stars to burn is gravity, the primal fuel is hydrogen, and in the case of our Sun, it consumes about 600 million tons of hydrogen every second. As this hydrogen is burned, or more accurately, "fused" into helium, photons of electromagnetic energy are released. The photons are eventually filtered through the Earth's atmosphere as solar energy. Thus solar energy is the result of a nuclear fusion process (not to be confused with the nuclear fission process of conventional nuclear reactors), without which there would not only be no life; there would be no fossil fuels; no wind; not even any uranium.

Given the critical need to make a fundamental transition from nonrenewable fossil fuel and nuclear-fission energy systems, it is only logical to consider the development of technologies that could utilize the relatively inexhaustible energy of the Sun. Because it is reasonable to assume that solar energy will someday serve as the primary energy source, it is important to understand the fundamental relationship that exists between solar energy and hydrogen.

*Primordial Hydrogen*

Excluding subatomic particles, modern physics provides a fairly clear picture of the origin of matter in the known universe, at least with respect to the development of the proton and the electron which are the basic components of the hydrogen atom. Hydrogen atoms, in turn, are the basic building blocks of all of the other 91 chemically distinct atoms that occur naturally.

The atomic number of an atom is equal to the number of protons (i.e., hydrogen nuclei) -- or electrons it contains. Thus hydrogen, with one proton and one electron, has an atomic number of 1. Carbon, on the other hand, has six protons and electrons, and therefore has an atomic number of 6. The proton has a positive electrical charge, and the electron has a negative charge, which explains their natural affinity for each other. These basic elements are believed to have been formed within the very first second of the origin of the universe when the "Big Bang" occurred some 15 billion years ago.

Were it not for gravity, which is the basic force that causes every particle of matter in the universe to be attracted to every other particle, the hydrogen atoms and other subatomic particles would have continued to fly away from each other from the initial force of the big bang. It was gravity, therefore, that was the primal force that caused the interstellar particles to concentrate in larger and larger masses. As the mass increases, so does the force of gravity, and eventually, the forces and pressures are sufficient for the interstellar clouds of hydrogen to collapse. When this collapse occurs, the hydrogen and other particles collide, and as a result, high enough temperatures (45 million degrees Fahrenheit) and pressures are created for the hydrogen to fuse into helium, at which point a star is born.

As the stars burn their supply of hydrogen fuel, four hydrogen nuclei are fused into a heavier helium nucleus. The heavier helium atoms form a dense, hotter core, and as the star reaches the point where it has consumed most of its primary hydrogen, it begins to burn -- or fuse -- the helium, converting it first to carbon and eventually to oxygen. Thus, a star is like the legendary Phoenix bird, destined to rise for a time from its own ashes of heavier elements. The more massive stars achieve higher central temperatures and pressures in their late evolutionary stages. As a result, when their helium is consumed, they are able to fuse the carbon and oxygen into still heavier atoms of neon, magnesium, silicon, and eventually silver and gold. Dr. Carl Sagan summarized this process in his book *Cosmos*:

> *"All the elements of the Earth except hydrogen and some helium have been cooked by a kind of stellar alchemy billions of years ago in stars....The nitrogen in our DNA, the calcium in our teeth, the iron in our blood, the carbon in our apple pies were all made in the interior of collapsing stars. We are made of starstuff."* [1]

From this physics and astronomical perspective, the relationship between solar energy and hydrogen is inseparable, in the sense that without hydrogen, there would be no Sun or other stars. In a similar context, the evolution of life on the Earth was, and is still is, a direct result of the dynamic interaction of these two primordial electrochemical elements.

### The Nanobes & Microbes

At present, no one knows exactly how the first living organisms evolved out of nonliving matter. What is known, however, is that atoms of hydrogen, carbon, oxygen and nitrogen eventually did evolve into an electrochemical structure of amino acids. These amino acids were, in turn, assembled into the complex architecture that makes up proteins, which includes the enzymes that are so critical to metabolism.  As a result, these building blocks of life may have been the first, and are certainly one of the most basic molecules of living organisms.

Since microbes such as bacteria, fungi and viruses operate on the micrometer scale (i.e., one-millionth of a meter), they have been referred to as microorganisms, or microbes.  But at the heart of the microbes are the enzymes and other proteins that operate on the nanometer scale (i.e., one-billionth of a meter).  This being the case, these protein-scale organisms may be thought of as "nanoorganisms," or "nanobes," and given their seemingly incredible array of activities, the nanobes could be viewed as a highly-advanced civilization that is more than 3.5 billion years old.

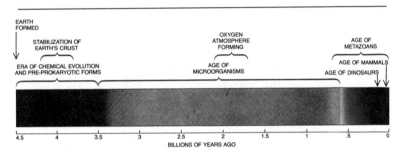

Figure 4.2: The evolution of life on Earth.
Note that the age of microorganisms (which are, in turn, made up of nanoorganisms) dominates the time span of biological evolution.  The illustration is from "Archaebacteria," by Carl R. Woese.  Copyright © June 1981 by *Scientific American, Inc.*  All rights reserved.

Although the human nervous system is a marvel of sophistication, it is literally the end product of billions of years of nanobial evolution, and it is directly operated and maintained by a current generation of nanobes that are at the heart of life itself.  Indeed, it would appear that nanobes make humans and other animals in the same way that humans make aircraft carriers.

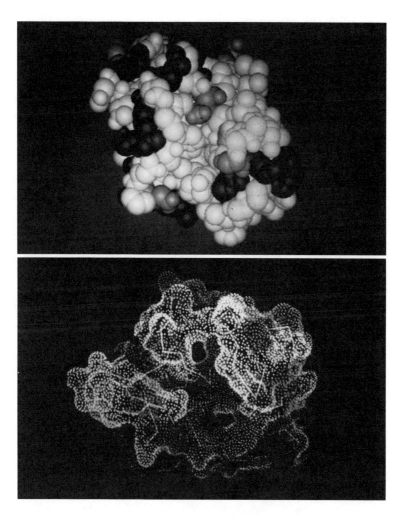

Figures 4.3 and 4.4: Nanobial Enzymes.
Enzymes and other proteins are at the heart of metabolism. The two images are of the enzyme *subtilisin*. The top image is of the molecular structure of atoms (excluding hydrogen), whereas the bottom image is a dot pattern of the enzyme's surface that was generated by a Cray supercomputer. Other enzymes have a completely different structure and appearance. Computer graphics modeling and photography by Arthur J. Olson, Ph.D., Research Institute of Scripps Clinic, Copyright © 1985.

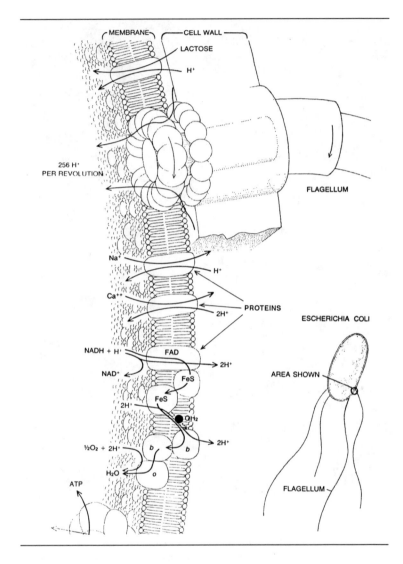

Figure 4.5: A microbial hydrogen-fueled engine.
Note that the flagellum (i.e., tail) of a typical bacteria is ac-
tually a molecular-scale rotary engine that requires 256
hydrogen nuclei (protons) per revolution. (From "How
Cells Make ATP," by Peter C. Hinkle and Richard E.
McCarty. Copyright © March 1978 by *Scientific
American, Inc.* All Rights Reserved.)

Just as an aircraft carrier is ultimately controlled by a group of specialized officers on the bridge, there are specialized nanobes that occupy and regulate the human command and control centers within the central nervous system. For example, Dr. Daniel L. Alkon, chief of the laboratory of molecular and cellular neurobiology at the National Institute of Neurological and Communicative Disorders and Stroke, has specialized in understanding the molecular mechanisms of memory. In a paper "Memory Storage and Neural Systems," published in *Scientific American*, Alkon explains that when individual neurons (i.e., brain cells) are in the process of learning, the flow of potassium ions through channels in the membranes is sharply reduced. The potassium-ion flow is what enables nerve cells to conduct electrical impulses, and when it is decreased, the cell becomes excited and electrical impulses can be triggered more readily. What is significant is that it is enzymes and other proteins that appear to regulate this critical ion flow[2].

Other nanobes are responsible for programming and maintaining the DNA biocomputer, which just happens to store the memory of amino acid sequences that code for all the different types of nanobial proteins that make up all living organisms. Nanobial proteins are the actual building blocks of life. Depending upon their configuration, they can become the structural material from which living tissue is made; they can become the hormones that regulate chemical behavior; or they can become the enzymes that mediate virtually all of the biochemical reactions in living organisms. Thus nanobes represent a critical link in the origin of life; they stand at the threshold of life where chemistry becomes biology. They are, in a biological sense, our creators.

Although nanobial enzyme reactions were first observed in the 1830's, little was known about their evolution prior to the 1970's. This was because the surface imprints of bacterial fossils provided little information about their origin or what their internal molecular structure could have been like. However, with the recent advances in molecular biology, scientists are no longer limited to just analyzing the geological fossil record. It is now known that by carefully analyzing the amino acid sequences within a cell's proteins and the nucleotide sequences of its nucleic acids (i.e., DNA and RNA), it is possible to provide a remarkably accurate picture of an organism's genetic and molecular origin. This living record provides far more information than geologic fossils because it provides a detailed picture of the evolution of molecular structures. Moreover, the living molecules reach back to a time long before the oldest fossils, to a period when the common ancestor of all life existed[3].

The primary objective and function of the initial primordial organisms was -- and still is -- to extract energy from other molecules. This was generally accomplished by rearranging the hydrogen atoms that bonded the various molecules together. This movement of hydrogen atoms has been referred to as *"transhydrogenation"* or fermentation, and such metabolic processes were then, and are now, the essence of what the biochemistry of life is all about. Thus Mother Nature has been successfully using the hydrogen energy system for several billion years.

The "technology" of transhydrogenation enabled the initial nanobe and microbe populations to increase rapidly, but they initially developed their hydrogen energy process around the nonrenewable hydrocarbon molecules they found in the primordial soup. This process went on for about 500 million years, at which point (roughly 3 billion years ago), their ever-increasing consumption of their nonrenewable resources began to exhaust their existing energy reserves. A distinguished Italian physicist, Dr. Cesare Marchetti, described their dilemma as remarkably similar to the current human global energy crisis. The nonrenewable primordial soup was being exponentially consumed in the same way as are the current global reserves of the nonrenewable fossil fuels[4].

*Photosynthesis*

Because the primordial soup was rapidly being exhausted, the nanobes were forced to modify their remarkably sophisticated molecular machines to utilize the vast and inexhaustible input of solar energy. The new molecule they developed was chlorophyll, and the relatively complex process it involves is referred to as *photosynthesis*. This innovative nanobial technology was one of the crucial links that allowed energy from the Sun to be incorporated into living, biological systems.

The eventual result of this nanobial reindustrialization effort, which allowed the nanobes to extract hydrogen from water, was the green plant. As a result, photosynthetic molecules represented a major technological breakthrough that allowed the nanobes to make a successful transition to renewable solar-hydrogen resources. Moreover, the problems of energy storage and transportation were also solved because the green plants could use the hydrogen liberated from water to produce carbohydrates and fats. As a result, the umbilical link with the primeval soup of non-renewable organic molecules was finally severed, and the nanobe's future was secure.

The scale and pace of evolution has increased enormously over the several billion years since the nanobes solved their energy crisis. But the basic molecular technology that they developed is still used today by their descendants, which occupy every living organism, including of course, human beings. Indeed, it has been estimated that 45 percent of the total intestinal gases generated in humans is made up of hydrogen[5]. Not surprisingly, there are also several examples in the scientific literature where this combustive mixture of gases has been ignited during surgical operations[6]. Indeed, in one case, an explosion even proved to be fatal to the unfortunate patient during a colonic polypectomy[7]. Hans Schlegel, a professor at the Institute for Microbiology of Gottingen University in Germany, explained the relationship of hydrogen as a source of energy for biological organisms:

> "All aerobic (i.e., requiring oxygen) organisms derive the energy necessary for the construction of their cell substance and to maintain their life functions from the reaction of hydrogen and oxygen. Man as well derives his metabolic energy through the slow combustion of hydrogen... although he is not being offered hydrogen in its gaseous state as nourishment, but rather as part of his foodstuffs [in which] it is weakly bonded to carbon." [8]

Table 4.1 provides a list of some of the most abundant elements found both in the universe and the human body. Note that the four elements that make up amino acids (i.e., hydrogen, carbon, nitrogen and oxygen) account for 99.9 percent of the elements in the universe and 98.7 percent of the elements that make up the human body.

Table 4.1:
Relative Abundance of Elements [9].

| Element | Universe | Human Body |
|---------|----------|------------|
| Hydrogen | 90.79% | 60.30% |
| Carbon | 9.08 | 10.50 |
| Nitrogen | 0.04 | 2.42 |
| Oxygen | 0.05 | 25.50 |
| Subtotal | 99.96% | 98.72% |
| Sodium | 0.00012% | 0.73% |
| Magnesium | 0.0023 | 0.01 |
| Aluminum | 0.00023 | |
| Silicon | 0.026 | 0.00091 |
| Phosphorus | 0.00034 | 0.134 |
| Sulfur | 0.00091 | 0.132 |
| Chlorine | 0.00044 | 0.032 |
| Potassium | 0.000018 | 0.036 |
| Calcium | 0.00017 | 0.226 |
| Iron | 0.0047 | 0.00059 |

From a biological perspective, each individual person is made up of some 100 trillion cells, each of which contains tens of thousands of nanobial organisms that still continue to use the basic process of hydrogen shuffling that was initially developed by their ancestors some 3.5 billion years ago. In any case, it should be clear that from both an astronomical or a biological perspective, the primordial relationships that exist between solar energy and hydrogen are symbiotic.

In much the same way, if one seriously proposes using solar energy to replace the use of fossil and nuclear fuels, the relationship between solar energy and hydrogen appears to be equally inseparable, because one cannot effectively work without the other. This is because the hydrogen needs to have some sort of a primary energy input (either solar or nuclear) in order to separate it from the other atoms of oxygen or carbon. Solar energy, on the other hand, will not be able to replace fossil or nuclear fueled energy systems unless it can be efficiently stored, transported and used as a combustion fuel in vehicles and power plants.

### The Water-Former

Hydrogen was first discovered in 1766 when the English chemist Henry Cavendish observed what he referred to as "inflammable air" rising from a zinc-sulfuric acid mixture. It was identified and named by the eighteenth century father of chemistry, Antoine Lavoisier, who showed that Cavendish's inflammable air did indeed burn in air to form water. He concluded it was a true element, and called it hydrogen, which is a Greek word that means "water former."

Hydrogen is the simplest, lightest and most abundant of the 92 regenerative elements in the universe. As an energy medium, it can never be exhausted because it recycles in a relatively short time. According to Abraham Lavi and Clarence Zener, two distinguished engineering professors at Carnegie-Mellon University (Pittsburgh, Pennsylvania), if hydrogen were used for electric power generation instead of fossil fuels, electricity costs could be reduced by as much as 50 percent [10]. The reasons are as follows:

1. Hydrogen can be burned in a combustion chamber rather than a more conventional boiler, thus high-pressure superheated steam can be generated and fed directly into a turbine. This technique reduces the capital cost of a power plant by 50 percent.

2. When hydrogen is burned, virtually no chemical pollution is generated. Thus, expensive pollution-control equipment, which can amount to one-third of the capital costs of conventional fossil fuel power plants is unnecessary.

3. The utilization of hydrogen fuel would allow power plants to be located in the vicinity of residential and commercial loads, thereby eliminating most of the power transmission costs and line losses. This would also allow waste heat to be efficiently utilized by residential and commercial buildings.

The fact that hydrogen burns cleanly and reacts completely with oxygen to produce water makes it a more desirable fuel than fossil fuels for virtually all industrial processes. One example would be the direct reduction of iron or copper ores by hydrogen rather than by coal in a blast furnace. Another example is that hydrogen could be used not only with conventional vented burners, but with unvented burners as well. This is important because nearly 30 to 40 percent of the combustion energy of conventional burners is vented as heat and combustion by-products.

*The Universal Fuel*

Hydrogen is not just another energy option like oil, coal, nuclear or solar, it is unique and stands alone because it is an inexhaustible "universal fuel" that can unite virtually all energy sources with all energy uses. Although solar technologies and resources have many attractive aspects, such as being renewable, modularized, and generally pollution-free, they also tend to have equally significant disadvantages, such as not being present in the right intensity at the right place at the right time. These are some of the major reasons why most solar technologies have not been economically competitive with nuclear or fossil-fueled facilities.

The hydrogen variable, on the other hand, fundamentally alters the global energy equation because it provides a realistic method of storing, transporting, and using the massive, but intermittent, supply of solar energy. As Figure 4.6 below indicates, hydrogen is a remarkable substance that is already used extensively in a wide-range of energy, industrial and chemical operations.

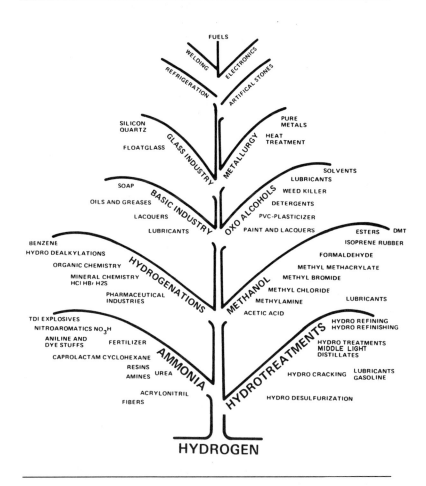

Figure 4.6: The Hydrogen Tree.
(Courtesy of Air Products and Chemicals, Allentown, Pennsylvania)

Hydrogen is a primary chemical feedstock in the production of fuels (including gasoline), lubricants, fertilizers, plastics, paints, detergents, electronics and pharmaceutical products. It is also a premium metallurgical refining agent and an important food preservative. Hydrogen can be extracted from a wide range of sources because it is in almost everything, from biological tissue and DNA, to petroleum, gasoline, paper, human waste, or water. It can be generated from nuclear plants, solar plants, wind plants, ocean thermal power plants or green plants.

In addition, hydrogen and electricity are directly complementary, in that one can be converted into the other. Thus hydrogen is a kind of energy currency that does not vary in quality with respect to origin or location. As a result, a molecule of hydrogen made by the electrolysis of water is identical to hydrogen manufactured from green plants (biomass), paper, coal gasification or natural gas facilities[11,12]. These are some of the reasons why the term "hydrogen economy" seems most appropriate.

### Primary vs. Secondary Energy Sources

Hydrogen is often referred to as a secondary energy carrier, rather than a primary energy source, because energy must be initially used to extract the hydrogen from water, natural gas, or other compounds that contain hydrogen. This classification is misleading because it assumes solar, coal, oil or uranium to be "primary" energy sources, implying that energy is not needed to extract them from the natural environment prior to their use.

Nuclear plants, for example, do not exist naturally. Indeed, they require vast energy expenditures and billions of dollars of investment before they can produce a single kilowatt of power. The same is true for the other so-called primary energy sources, including coal, oil and natural gas. Although coal and natural gas come closest to being true primary energy sources that can be burned directly with little or no refining, energy must still be required to extract these "natural" energy sources and deliver them to the place the energy is needed. Even if drilling for oil were not required, energy must still be used to refine it into a useable energy form such as gasoline or diesel fuel. The late R. Buckminster Fuller provided an interesting perspective regarding the true cost of a barrel of oil in an article, "Geoview," published in *World Magazine* in 1973:

> *"Scientific calculation shows that the amount of time and energy invested by nature to produce one gallon of petroleum 'safely deposited' in subterranean oil wells, when calculated in foot-pounds of work and chemical time converted to kilowatt hours and the present commercial rates at which electricity is sold amounts to approximately one million dollars per gallon..."* [13]

Moreover, current energy cost accounting methods do not take into account the many global environmental problems that are a direct result of finding, transporting and burning the fossil fuels. In sharp contrast, when hydrogen is used as a fuel, its combustion by-product is essentially water vapor. If hydrogen is burned in air, which is mostly nitrogen, oxides of nitrogen can be formed as they are in gasoline-fueled engines. But these undesirable by-products can to a large extent be eliminated in hydrogen-fueled engines by lowering the combustion temperatures of the engine (for example, by injecting water into the cylinder). As a result, hydrogen is the cleanest-burning combustion fuel that could be used for transportation and other industrial applications.

Laboratory tests have shown that the air coming out of a hydrogen-fueled engine was actually cleaner than the air that entered the engine[14]. Accordingly, if all of the automobiles, trucks, buses, aircraft, and trains were using gaseous or liquid hydrogen fuel, the air pollution in the major urban areas would be, to a large extent, eliminated. In addition, the especially serious problems of acid-rain, stratospheric ozone depletion, and carbon dioxide accumulations could all be dramatically impacted by the use of hydrogen for automotive and industrial applications. Thus, an industrial transition to a hydrogen energy system may be the only realistic solution to resolving the potentially catastrophic environmental problems which threaten the biological life support systems of the Earth.

This, in and of itself, is reason enough to develop and deploy solar-hydrogen energy systems.

## Hydrogen Storage Systems

Once hydrogen is produced, it must be stored as either a gas, a liquid, or as a solid metal, polymer or liquid hydride (such as gasoline or methanol). Numerous studies have determined that for large-scale storage, the preferred method would be to store gaseous hydrogen underground in aquifers, depleted petroleum or natural gas reservoirs or man-made caverns that have resulting from mining operations. However, one of the primary obstacles that has made it difficult to use hydrogen as an automotive fuel is storing it safely and efficiently on-board aircraft and ground vehicles. Although it is possible to store hydrogen as a high pressure gas in steel containers, this method is not practical for automotive vehicles because of the weight of the storage cannisters and the fact that they would pose a serious safety hazard in the event of a collision. This leaves the other options of storing hydrogen as solid or liquid hydride, a low-temperature cryogenic liquid, or as a combination of the two.

## Hydrogen Hydrides

Hydride materials absorb hydrogen like a sponge, and then release it, usually when heat is applied. While there are hundreds of potential hydride material selections, the most common hydride systems that have thus far been used in automotive vehicles have consisted of metal particles of iron and titanium that were initially developed by researchers at Brookhaven National Laboratory. In tests conducted by American investigators at Billings Energy Corporation (Provo, Utah) and German investigators at Daimler-Benz (Stuttgart, Germany), these hydride systems have been shown to be a very safe method of storing hydrogen in automobiles, but they are about five times heavier than liquid hydrogen storage systems.

There are many other hydride options under development that do not have such severe weight penalties, such as magnesium-nickel alloys, non-metallic polymers, or even liquid hydride systems that are able to use engine heat to disassociate fuels like methanol into a relatively clean burning mixture of hydrogen and carbon monoxide. However, with the present iron-titanium hydride systems, if a typical range of 500 kilometers (310 miles) is to be provided, the storage system would weigh about 2,600 kilograms (5,700 pounds). In contrast, a liquid hydrogen tank providing a similar range would weigh about 136 kilograms (300 pounds), while a comparable gasoline tank would weigh about 63 kilograms (138 pounds)[15].

Figure 4.7: Iron-titanium hydrogen hydrides.
Such hydride alloys are milled into the small gray particles that are pictured above in the cutaway of the hand-held hydride storage tank.
(Courtesy of the Tappan Appliance Company)

Figure 4.8: Dr. Helmut Buchner, Daimler-Benz hydrogen-project leader, displays a model of a hydride storage system used in the vehicle behind him.

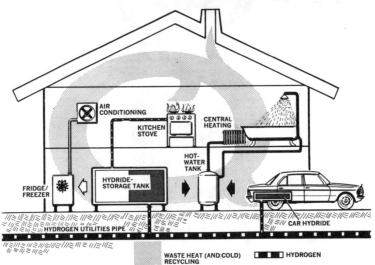

Figure 4.9: Hydrogen residential delivery system. (Both photographs reprinted from *Popular Science* with permission © 1978 *Times Mirror Magazines, Inc.*)

If an electric vehicle with existing lead-acid batteries were to have a 500 kilometer (310 mile) range, the weight of the batteries would be around 3,000 kilograms (6,600 pounds). More efficient automotive battery systems are being developed by numerous automotive manufacturers, but the most efficient electric vehicles in the future may not be energized by batteries at all -- but by "fuel cell" systems that are able to convert hydrogen and oxygen directly into electricity. Such systems, however, would be predicated on having hydrogen fuel readily available.

Liquid Hydrogen (LH$_2$)

In order for hydrogen gas to be liquefied, it needs to be cooled to a very low temperature: minus 421.6 degrees Fahrenheit, which makes liquid hydrogen a cryogenic fuel. Cryogenics, the study of low temperature physics, is a rapidly expanding area of research in both engineering and biology. If, for example, a beaker of liquid hydrogen were sitting on a table at room temperature, it would be boiling as if it were water sitting on a hot stove. If the beaker of liquid hydrogen were spilled on the floor, it would be vaporized and dissipated in a matter of seconds.

If liquid hydrogen were poured on one's hand, it would would only feel slightly cool to the touch as it would slide through one's fingers and fall to the ground. This is due to a thermal barrier that is provided by the skin. On the other hand, if one placed one's hand in a vessel containing liquid hydrogen, severe injury would occur in seconds because of the ultra-cold temperatures. However, such a circumstance would be unlikely in typical automotive environments where the main concern involves refueling and collisions. In most aircraft and automotive accidents, the most serious concern is the possibility of a fuel-fed fire and/or explosion. Under such circumstances, liquid hydrogen would generally be a preferred fuel.

Liquid hydrogen is one of the most viable storage options that could be utilized on a large scale in the transportation sector because it most resembles gasoline or aviation kerosene in terms of space and weight. Although a liquid hydrogen storage tank for an automobile is expected to be about five times heavier (dry weight) than a typical 30-pound gasoline tank, in larger vehicles that carry greater volumes of fuel, such as trucks, trains or aircraft, the difference in tank weight could be more than offset by the difference in fuel weight.  For example, Lockheed studies have shown that a large commercial aircraft would have its overall takeoff weight reduced by as much as 40 percent if liquid hydrogen were used instead of conventional aviation kerosene[16].

One of the most important aspects of having hydrogen fuel on board a vehicle is that it would then be possible to have a small and highly efficient fuel cell-Stirling engine cryocooler system that could provide air conditioning without the use of ozone-destroying chlorofluorocarbon (CFC) Freon gases.  Other advantages of using liquid hydrogen include the following:

* $LH_2$ has the lowest weight per unit of energy.
* $LH_2$ has relatively simple supply logistics.
* $LH_2$ has normal refuel time requirements.
* $LH_2$ is generally safer than gasoline in accidents.

Figure 4.10: An artist's concept of a liquid hydrogen-fuel and water system in a test vehicle.
(Courtesy of Los Alamos National Laboratory)

Figure 4.11: A liquid hydrogen-fueled Buick modified by investigators at the Los Alamos National Laboratory.

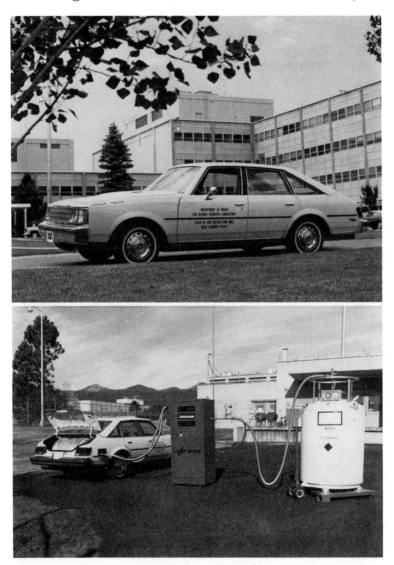

Figure 4.12: A liquid hydrogen self-service refueling pump and cryogenic fuel storage tank. (Courtesy of German Aerospace Research Establishment, Stuttgart, and the Los Alamos National Laboratory, Los Alamos, New Mexico)

Figure 4.13: A BMW Liquid-Hydrogen Fueled Automobile.
This is the first liquid hydrogen vehicle modified by a
major automobile manufacturer.

Figure 4.14: The liquid hydrogen cryogenic storage tank
mounted in a well insulated trunk of the BMW sedan.
(Courtesy of BMW of North America, Inc.)

Figure 4.15: Fuselage cross-section of conventional kerosene and liquid hydrogen-fueled aircraft [17].

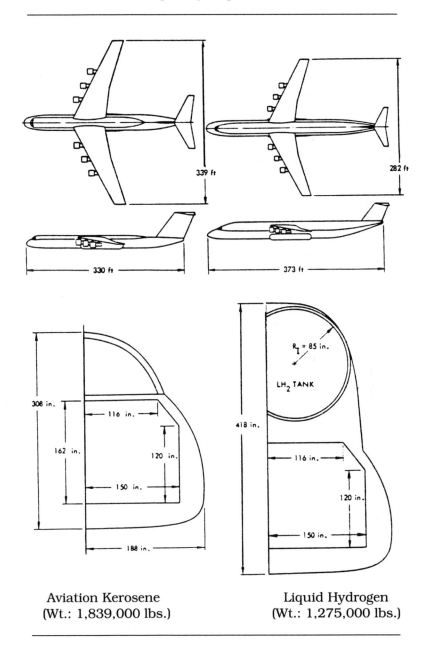

Aviation Kerosene
(Wt.: 1,839,000 lbs.)

Liquid Hydrogen
(Wt.: 1,275,000 lbs.)

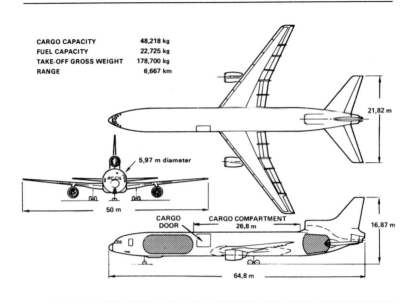

CARGO CAPACITY          48,218 kg
FUEL CAPACITY           22,725 kg
TAKE-OFF GROSS WEIGHT   178,700 kg
RANGE                   6,667 km

21,82 m

5,97 m diameter

50 m

CARGO DOOR     CARGO COMPARTMENT 26,8 m     16,87 m

64,8 m

Figure 4.16:  Overall dimensions and characteristics of
a liquid hydrogen-fueled aircraft proposed by
Lockheed Aircraft Corporation [18].
(Illustration courtesy of Lockheed Missiles & Space Co.)

*Liquid Hydrogen Disadvantages*

Although liquid hydrogen offers many advantages in
vehicular applications, it also has several significant disadvantages that need to be put into perspective.  Ultra-
cold cryogenic fuels like liquid hydrogen are more difficult
to handle and substantially more difficult to store, in comparison with hydrocarbon fuels like gasoline or aviation
kerosene.  With the current generation of highly-insulated
double-walled, vacuum-jacketed storage tanks manufactured by the German Aerospace Research Establishment
DFVLR (Stuttgart), the liquid hydrogen will evaporate at a
rate of about 8 percent per day.  Because the evaporation
increases the pressure on the tank wall, the gaseous hy-

drogen must be vented to the atmosphere to keep the tank from rupturing. Investigators at Los Alamos National Laboratory found that in a 1979 liquid hydrogen-fueled Buick they were testing, a full tank of liquid hydrogen would evaporate in about 10 days[19].

This venting of the fuel not only presents the problem of an empty fuel tank, but if the vehicle is in an enclosed space such as a garage, the vented hydrogen could pose a serious risk of being ignited due to hydrogen's wide flammability limits. Although hydrogen explosions are rare, any combustible gas in an enclosed space is an obvious safety concern. As an interim solution to this problem, the engineers at DFVLR installed a small burner underneath a liquid hydrogen-fueled BMW test vehicle. The burner is continuously ready for operation and, upon reaction of the pressure release valve, a pilot flame safely burns the escaping hydrogen.

However, according to a technical report by DFVLR, "Liquid Hydrogen as a Vehicular Fuel," stationary liquid hydrogen storage tanks that are used in laboratories are able to keep the hydrogen in a liquid state for several months. Thus it should be possible to engineer vehicular storage tanks that would maintain hydrogen in a liquid state for several weeks[20]. The small quantity of hydrogen evaporating from such tanks could also be directed to a fuel cell that could use the hydrogen to generate electricity which could then recharge the battery, thereby eliminating the need for the engine alternator. It is also possible to vent the vaporized hydrogen gas to a supplemental hydride system for storage, and as hydrogen is integrated as a primary fuel, such a multiple storage option will surely be refined. Other disadvantages of liquid hydrogen include the following:

1. Double-walled vacuum-jacketed storage tanks and piping that are required for liquid hydrogen are more expensive than conventional fuel storage tanks. A gasoline tank for an average automobile is about $150, whereas a liquid hydrogen storage tank for the same automobile would be expected to cost from $1,000 to

$2,000, assuming production volumes of 100,000 units per year[21].

2. Because of the energy density of liquid hydrogen, it requires a fuel tank to be roughly three to four times as large in volume as its gasoline or aviation fuel counterpart.

3. Liquid hydrogen fuel systems would require changes in the entire energy infrastructure, such as pipelines, transport vehicles and end-use systems, such as stoves, automotive engines and fuel pumping systems.

While these disadvantages of using liquid hydrogen are substantial, they can be minimized, and more importantly, they need to be put into perspective. For example, it may seem unreasonable to spend one or two thousand dollars for a liquid hydrogen storage tank, until one realizes that the current emissions control equipment required on gasoline-fueled engines (which would not be required on hydrogen-fueled engines) costs about $1,000. In addition, if liquid hydrogen were used on all new automobiles in the U.S., production volumes of cryogenic storage tanks would not be in the thousands, but in the millions. As such, the cost of the individual cryogenic tanks would be expected to drop below $1,000.

Although a liquid hydrogen fuel tank may cost substantially more than a gasoline tank, anyone who has seen what happens to people who have had the unfortunate experience of being burned from a gasoline fire would never argue about the additional costs of a liquid hydrogen fuel tank. Although cryogenic fuels are somewhat more difficult to handle, a self-service liquid hydrogen pumping station has already been developed by DFVLR, that is pictured in Figure 4.12. After using this system, investigators at Los Alamos National Laboratory concluded the following:

*"...liquid hydrogen storage and refueling of a vehicle can be accomplished over an extended period of time without any major difficulty."* [22]

Although a liquid hydrogen fuel tank will be three or four times larger than a conventional fuel tank, it will require only minimal design changes for most vehicles. Even in the case of aircraft, the changes are not significant. In order to minimize the amount of drag that would result from having larger wing tanks, Lockheed engineers have proposed extending the length of the fuselage of the aircraft to accommodate the larger liquid hydrogen tanks as shown in Figure 4.16. Other configurations under study by the U.S. Defense Department are shown in Figure 4.15. With respect to the increased costs associated with making a changeover to hydrogen energy systems, it needs to be remembered that the environmental costs of finding, transporting and burning fossil fuels are not calculated in the current energy pricing structure. This is in spite of the fact that the increasing concentrations of atmospheric pollution are costing billions of dollars in additional health care costs, crop losses, and the corrosion of buildings and other structures.

### The Hydrogen Engine

Various automotive research engineering teams in Germany, the U.S., Japan, France, the Netherlands, Brazil and the Soviet Union are involved in hydrogen research and development. Hydrogen-fueled engines have been shown to be more energy efficient because of their complete combustion. Moreover, gasoline and diesel-fueled engines form carbon deposits and acids that erode the interior surfaces of the engine and contaminate the engine oil. This, in turn, increases wear and corrosion of the bearing surfaces. Since hydrogen-fueled engines produce no carbon deposits or acids, it is anticipated they will require considerably less maintenance. In addition, hydrogen fuel can also be utilized with more efficient Stirling cycle engines or fuel cells that could allow electric vehicles to be truly practical.

Automotive engineers have been aware of hydrogen's favorable combustion characteristics since the early 1900's. A German engineer, Rudolf A. Erren, began optimizing internal combustion engines to use hydrogen in the

1920's, and he is generally recognized by the hydrogen technical community as the father of the hydrogen-fueled engine. Erren modified many trucks and buses, and Allied forces even captured a German submarine in World War II that had not only a hydrogen-fueled engine, but hydrogen-powered torpedoes as well that were initially designed and patented by Erren and his associates.

Figure 4.17: Rudolf A. Erren, the father of the hydrogen engine. (Photo was taken in Germany in 1979 and is reprinted with permission from Peter Hoffmann, *The Forever Fuel, The Story of Hydrogen*)

Unfortunately, hydrogen vehicle modification projects in the U.S. have not been seriously undertaken by any of the major automotive manufacturers, who have generally viewed such research as too speculative. Rather, it has been small research and development efforts at various universities, government laboratories or small research and development companies, such as Billings Energy Corporation, that have taken the lead in developing innovative hydrogen-fueled vehicles.

*Roger Billings*

The first hydrogen-fueled automobile in the U.S. was a Model A Ford truck, developed in 1966 by Roger Billings. At the time, Billings was still a student in high school, but several years later as as an undergraduate student in chemistry at Brigham Young University, he won a 1972 Urban Vehicle Design Contest with a hydrogen-fueled Volkswagen. Billings eventually established Billings Energy Corporation (Provo, Utah) and went on to modify a wide range of automotive vehicles, including a Winnebago motor home that not only had the engine fueled by hydrogen, but the electrical generator and all of the vehicle's appliances as well.

Figure 4.18: Roger Billings.
(Courtesy of American Academy of Science)

Billings went on to design and construct the world's first "hydrogen homestead," pictured in Figure 4.19, which had virtually all of the homes appliances modified to operate on hydrogen. Shown in front of the Billing's hydrogen home is a dual-fueled (i.e., hydrogen or gasoline) Cadillac

Seville and a Jacobsen farm tractor. The driver of the dual-fueled vehicle is able to change from hydrogen fuel to gasoline while driving with the flip of a switch from inside the vehicle. As a result of his many years of research, Billings and his colleagues have been able to demonstrate that hydrogen could indeed be used as a universal fuel in end-use applications.

Figure 4.19: The Billings Hydrogen Homestead.
(Courtesy of the Tappan Appliance Company)

The photos in Figure 4.20 show a special burner head that had been modified by the Tappan Company for hydrogen combustion. Because hydrogen burns with an invisible flame, it was necessary to utilize a material that could allow someone to know the range was on. The engineers at Tappan utilized a steel wool catalyst that rests on the burner head to solve this problem. The steel wool is covered by a stainless steel shroud (shown to the left of the steel wool) that slips over the catalyst and burner head to give a clean, uncluttered appearance pictured in the bottom photo. The stainless steel mesh glows when it is heated, resembling an electric range surface when the burner is on.

Figure 4.20: A Tappan Hydrogen-Fueled Gas Range.
(Courtesy of the Tappan Appliance Company)

To demonstrate that even small portable appliances can be modified to use hydrogen fuel, Billings Energy Corporation also adapted a Coleman Stove, pictured in Figure 4.21, for hydrogen combustion. The small hydrogen storage tank on the right utilizes iron-titanium metal hydrides.

Figure 4.21: A Hydrogen-Fueled Coleman Stove.
(Courtesy of American Academy of Science)

Because of anticipated fossil fuel shortages, hydrogen research programs have been undertaken by the U.S. Air Force, Navy and the Army since the 1940's (refer to Figure 4.22). But because of the difficulties of storing hydrogen on board vehicles, and the fact that it was more expensive to produce than hydrocarbon fuels refined from oil, meant that the hydrogen fuel option was not pursued. The Department of Defense was not concerned with environmental considerations, and since it has to purchase fuel on the open market like every other consumer, fuel costs were the primary concern. Prior to the Arab oil embargo in 1973, oil was selling for less than three dollars per barrel, and as long as there was not an emergency, the fuel supply problem was essentially left to the oil companies to worry about.

Figure 4.22: Comparisons of a M-60 tank fueled
with diesel fuel, in contrast to liquid
hydrogen [23].

---

Diesel / Length: 273 feet / Weight: 114,000 lbs.

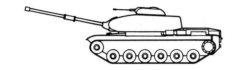

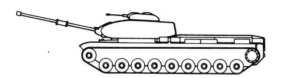

Liquid Hydrogen / Length: 368 feet / Weight: 130,171 lbs.

---

The hydrogen that has been manufactured for industry is principally made by reacting natural gas with high temperature steam, thereby separating the hydrogen from the carbon. But manufacturing hydrogen from fossil fuel resources does not solve the fossil fuel depletion problem. Making hydrogen from water through electrolysis was initially anticipated by nuclear engineers who believed that nuclear-generated electricity would be inexpensive enough to make hydrogen, but the high cost and unreliable nature of commercial reactors effectively eliminated the nuclear option, and thus, the hydrogen economy was put on hold. But after the Arab oil embargo in 1973, there were long gas lines in the U.S. and the price of oil quadrupled. This prompted renewed research into alternative energy systems, including renewable solar technologies. As a result, significant progress has been made in identifying renewable energy systems that could be utilized for large-scale hydrogen production. Some of the most cost effective solar technologies developed thus far will be discussed at length in Chapter 5.

*Hydrogen Safety*

Most people tend to assume that hydrogen is especially dangerous. There are many reasons for this belief. Some people think hydrogen energy is somehow related to the "hydrogen" bomb. This is perhaps understandable because of the common usage of the word hydrogen, but hydrogen used as a fuel involves a simple chemical reaction that involves the transfer of electrons, whereas the hydrogen bomb involves a high-temperature nuclear fusion reaction similar to that which occurs in stars.

Other people remember that the great German airship *Hindenburg* was using hydrogen as a lifting-gas when it was destroyed by fire in a spectacular crash in 1937 that was captured on film. This event had a profound emotional impact on a great many people, but while 35 people died in the unfortunate accident, it is hardly ever mentioned that 62 other people survived. It is also rarely mentioned that prior to its fatal crash in 1937, the *Hindenburg* had successfully completed 10 round trips between the U.S. and Europe, and its sister ship, the *Graf Zeppelin*, had made regular scheduled transatlantic crossings from 1928 to 1939 with no mishaps. Indeed, of the 161 rigid airships built and flown between 1897 and 1940, nearly all of which used hydrogen as a lifting gas, only 20 were destroyed by fires. Of the 20, seventeen were lost in military incidents that in many cases resulted from hostile enemy fire during World War I. That is an excellent safety record for the technology of the day.

Hydrogen does have a wider range of flammability when compared to gasoline. For example, a mixture as low as 4 percent hydrogen in air, or as high as 74 percent will burn, whereas the fuel to air ratios for gasoline are only from 1 to 7.6 percent. In addition, it takes very little energy to ignite a hydrogen flame, about 20 micro-joules, compared to gasoline which requires 240 micro-joules. However, the hazardous characteristics of hydrogen are offset by the fact that it is the lightest of all elements, which means it has a very small specific gravity. Because the diffusion rate of a gas is inversely proportional to the square root of its specific gravity, the period of time in which hydrogen and oxygen are in a combustible mixture is much shorter than for other hydrocarbon fuels. The lighter the element is, the more rapidly it will disperse if it is released in the atmosphere.

In the event of a crash or accident where hydrogen is released, it would rapidly disperse up and away from people and other combustible material within the vehicle. Gasoline and similar hydrocarbon fuels on the other hand are heavy because the hydrogen is bonded to carbon. As a result, when the hydrocarbon fuels vaporize, their gases will sink rather than rise in the atmosphere. This means burning gasoline falls on people and literally burns them alive. That is why in the relative world, hydrogen would, in most cases, be a more desirable vehicular fuel if a serious accident were to occur.

It is important to put the *Hindenburg* accident into perspective. On March 27, 1977, two fully-loaded Boeing 747 commercial aircraft crashed into each other on a foggy runway in the Canary Islands. This disaster, the worst in aviation history, took 583 lives, compared to the 35 lost in the crash of the *Hindenburg*. Investigators concluded many of the deaths in the Canary Islands accident were a result of the kerosene-fueled fire that *raged* for more than 10 hours. G. Daniel Brewer, who was at the time the hydrogen program manager for Lockheed, told an audience of experts six weeks later that if both aircraft had been using liquid hydrogen fuel instead of kerosene, hundreds of lives could have been saved for the following reasons:

1. Liquid hydrogen cannot react with oxygen and burn until it first vaporizes into a gas. As it does evaporate, it dissipates rapidly as it is released in open air. As a result, the fuel-fed portion of the fire would have only lasted a few minutes; not many hours as with conventional liquid hydrocarbon fuels.

2. With hydrogen, the fire would have been confined to a relatively small area because the liquid hydrogen would rapidly vaporize and disperse in the air, burning upward, rather than spreading like kerosene.

3. Heat radiated from the hydrogen fire would be significantly less than that generated by a hydrocarbon fire. As a result, only persons and structures immediately adjacent to the flames would be affected.

4. The hydrogen fire would produce no smoke or toxic fumes, which can in many cases be even more dangerous to the passengers and crew than the actual flames themselves.

5. With liquid hydrogen fuel storage tanks, the gaseous hydrogen that vaporizes will fill the empty volume inside the tanks. The hydrogen is not combustible because no oxygen is present. With conventional gasoline or other hydrocarbon fuel tanks, air fills the empty volume of the tanks and combines with vapors from the fuel to create a potentially combustible mixture that is a formidable safety hazard [24].

These are some of the main reasons why liquid hydrogen would be relatively safe in the event of an accident, in comparison to conventional hydrocarbon fuels.

Figure 4.23 provides an artist's conception of a typical air terminal in a liquid hydrogen airport. A hydrogen-fueled Lockheed L-1011 commercial aircraft is in the process of being refueled by a ground crew. In the distance, there are two vacuum-jacketed liquid hydrogen storage spheres (about 35 feet in diameter). Vacuum-jacketed cryogenic fuel lines contained in a trench covered with an open steel grate carry the liquid hydrogen from the storage vessels to a series of hydrant pits, like the one being used to refuel the aircraft. A "cherry picker" truck makes the connection between the hydrant pit and the liquid hydrogen-fueled aircraft. Of the two lines shown, one taps off gaseous hydrogen displaced from the aircraft tanks by the incoming liquid hydrogen and returns it to the liquefaction plant.

Lockheed studies have concluded that in addition to hydrogen's favorable safety characteristics, liquid hydrogen-fueled aircraft will have other significant advantages over their fossil-fueled counterparts; they will be lighter; quieter; will require smaller wing areas; shorter runways; and will minimize pollution. They will also use less energy for two reasons: less hydrogen fuel is needed per flight mile; and less energy is required to manufacture the hydrogen fuel compared with alternative fossil fuels. Because liquid hydrogen has the highest energy content per weight of fuel, the range of an aircraft could be roughly doubled, even though its takeoff weight would remain essentially the same.

Figure 4.23: An artist's illustration of a liquid hydrogen-fueled commercial airport.
(Courtesy Lockheed Aircraft Corporation[25])

*Hydrogen Explosions*

The *Hindenburg* did not explode, as most people assume. Rather, it caught fire, and as the flames rapidly spread, the airship sank to the ground. Although sabotage was suspected, most experts believe the fire was started because the airship was venting some of its hydrogen (in order to get closer to the ground) in the middle of an electrical thunderstorm. In addition, the airship was simultaneously moored to the ground by a steel cable, which only increased the risk of the hydrogen being ignited by static electrical discharge. Hydrogen explosions are extremely powerful when they occur, but they are extremely rare. This is because hydrogen needs to be in a confined space for an explosion to occur. Out in the open, it is almost impossible to bring hydrogen to an explosion without the use of a heavy blasting cap [26].

In 1974, Paul M. Ordin, a research analyst working for NASA presented a paper to the *Ninth Intersociety Energy Conversion Engineering Conference*, which reviewed 96 accidents or incidents involving hydrogen. NASA tanker trailers had transported more than 16 million gallons of liquid hydrogen for the Apollo-Saturn program alone, and while most mishaps were of a highly specialized nature, there were five serious highway accidents that involved extensive damage to the liquid hydrogen transport vehicles. These accidents were such that if conventional gasoline or kerosene had been involved, a spectacular blaze would have been expected, but due to the physical characteristics of liquid hydrogen, none of the accidents resulted in either a hydrogen explosion or fire [27].

The U.S. Defense Department has been conducting liquid hydrogen-fuel safety research since 1943. In tests undertaken by the Air Force Flight Dynamics Laboratory at Wright-Patterson Air Force Base, armor-piercing incendiary and fragment-simulator bullets were fired into aluminum storage tanks containing both kerosene and liquid hydrogen. The test results indicated that the liquid hydrogen was safer than conventional aviation kerosene in terms of gross response to a ballistic impact [28].

Other tests involved simulated lightning strikes, using a 6-million volt generator that shot electrical arcs directly into the liquid hydrogen containers. In neither of these tests nor the earlier Defense Department tests did the liquid hydrogen explode. Fires did ignite as a result of the simulated lightning strikes, but in the case of liquid hydrogen, the fires were less severe even though the total heat content of the hydrogen was twice that of kerosene.

These are some of the main reasons why liquid hydrogen would be more desirable than fossil-based fuels in combat situations where an aircraft's fuel tanks could be penetrated by explosive bullets or fragments.

One example of a dangerous situation where explosive mixtures of hydrogen and oxygen were present in a confined space occurred when the highly-publicized partial meltdown happened in 1979 at the Three Mile Island (TMI) nuclear facility in Pennsylvania. Nuclear reactors operate at very high temperatures, and the only thing that prevents their six to eight inch thick steel reactor vessels from melting is the large amounts of cooling water that must be continuously circulated in and around the reactor vessel.

An average commercial-sized reactor requires about 350,000 gallons of water per minute. During the process of nuclear fission, the center of the uranium fuel pellets in the fuel rods heat up to about 4,000 degrees Fahrenheit. The cooling water keeps the surface temperature of the pellets down to about 600 degrees, but if, for any reason, the water is not present, within 30 seconds the temperatures in the reactor vessel will soar to over 5,000 degrees. Such a searing temperature is not only high enough to melt steel, it will also thermochemically split any water present into a highly explosive mixture of hydrogen and oxygen -- which is what happened at TMI. If a spark had ignited the hydrogen gas bubble that drifted to the top of the containment building at TMI, the resulting powerful explosion could have fractured the containment building. This, in turn, would have resulted in the release of large concentrations of radiation at ground level.

Fortunately, the hydrogen gas bubble at TMI was successfully vented, but as long as it remained in the confined space of the containment building, its potential for detonation was of paramount concern. It should be noted that a hydrogen gas bubble developing from a serious nuclear reactor accident is a highly unusual event, even for the current generation of nuclear-fission reactors. It is an example of the specialized conditions that are required for hydrogen to explode. Such an isolated accident is essentially unrelated to the safety issue of using gaseous or liquid hydrogen as a vehicular fuel.

*NASA*

Dr. Warner Von Braun, a German rocket engineer who helped to develop the V-2 rockets in World War II, directed the first engineering effort to utilize liquid hydrogen as a rocket fuel. After the war, Von Braun was responsible for helping to direct the U.S. space program, which eventually evolved into the National Aeronautics and Space Administration (NASA). Because liquid hydrogen has the greatest energy content per unit weight of any fuel, NASA engineers selected liquid hydrogen as the primary fuel for the Saturn 5 moon rockets and the Space Shuttle. NASA also funded research by numerous aerospace firms, including Lockheed and Boeing, to determine if liquid hydrogen could be economically used in conventional commercial aircraft, and what modifications would need to be made to airports and related fuel systems if liquid hydrogen were to be used as a commercial aircraft fuel[29].

Figure 4.24: The Space Shuttle. Note the clean exhaust from the shuttle's main hydrogen-fueled engines compared to the two solid-fueled rocket bosters. (Courtesy of NASA)

The Space Shuttle in Figure 4.24 shows the main liquid hydrogen-oxygen tank, which is the largest of the three external tanks. The two smaller boosters utilize a solid-aluminum based fuel. NASA has used and handled large quantities of gaseous and liquid hydrogen for many years, which involved developing the necessary pipelines, storage tanks, barges and transport vehicles. As a result of this extensive field experience, officials at NASA have been able to conclude that hydrogen can be as safe, and is in some ways safer, than gasoline or conventional aviation hydrocarbon fuels.

Although NASA's design engineers initially wanted to develop a reusable manned liquid hydrogen-fueled launch vehicle for the space shuttle program (pictured in Figures 4.25, through 4.28), the members of Congress refused to appropriate the additional funds that would be necessary. As a result, the less expensive solid rocket boosters were used instead. This turned out to be a tragic mistake when one of the seals of the solid rocket boosters failed due to a cold weather launch. This resulted in the explosion of the *Challenger* shuttle in 1986 and the loss of its entire crew, including the first teacher ever to attempt a spaceflight.

Figure 4.25: NASA engineers initially planned to utilize a liquid hydrogen-fueled manned shuttle launch vehicle pictured above, but budget cuts forced the use of solid-rocket boosters instead. (Courtesy of NASA)

Figure 4.26:  A projected mission profile for both the manned shuttle launch vehicle and an early version of the space shuttle itself.

Figure 4.27: Separation of space shuttle from manned launch vehicle.  (Illustrations courtesy of NASA)

Figure 4.28: An artist's concept of the shuttle's manned
launch vehicle returning to Earth for a landing.
(Illustration courtesy of NASA)

*Hypersonic Aerospacecraft*

While liquid hydrogen fuel offers many advantages for
subsonic aircraft, its advantages increase substantially for
supersonic, and the much more advanced aircraft that
have been referred to as "hypersonic aerospacecraft."
Such advanced vehicles are now under development in the
U.S., Germany and Great Britain. In the U.S., numerous
governmental agencies, including NASA, the Navy, and the
Air Force have all been actively researching such advanced
aircraft for many years. When operational, such advanced
transport vehicles would make the current generation of
space shuttles essentially obsolete. The hypersonic
aircraft pictured in Figure 4.29 was initially designed by
Lockheed in the 1970's as an advanced supersonic
transport (SST). Such a vehicle was being designed to
carry about 200 passengers at speeds in excess of Mach 6
(i.e., about 4,000 miles per hour).

Figure 4.29: An artist's concept of a 4,000 mph liquid
hydrogen-fueled hypersonic aircraft.
(Courtesy of Lockheed Missiles & Space Corporation)

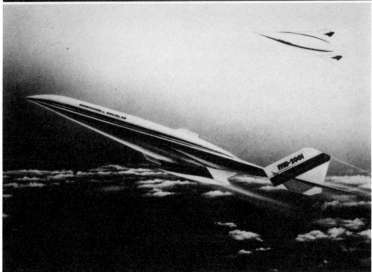

Figure 4.30: The McDonnell Douglas 8,000 mph X-30
liquid hydrogen-fueled National AeroSpace Plane.
(Courtesy of McDonnell Douglas Corporation)

Figure 4.30 provides a more updated version of an aerospace craft, referred to as the X-30 (experimental) National Aerospace Plane (NASP). Such an aircraft could travel from New York to Los Angeles in about 12 minutes, at speeds in excess of 8,000 miles per hour. Although commercial interest in the SST program did not materialize in the U.S. in the 1960's, the U.S. Air Force continued funding advanced SST and hypersonic research because of the obvious military implications of such technology. As a result of this ongoing research, by the time the Space Shuttle *Challenger* exploded some 25 years later, aerospace engineers in both the U.S. and Great Britain (and perhaps the Soviet Union) had made enough technical progress in supercomputers and propulsion systems to move from the research stage to actually fabricating test systems. Hypersonic aircraft will be able to take off from conventional runways, achieve speeds in excess of Mach 15, ascend into Earth orbit and fly back to the Earth in powered flight. While such capabilities would have important military advantages, such spacecraft would also accelerate the commercialization of space.

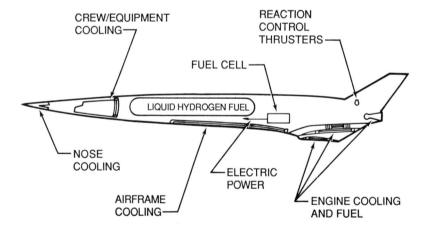

Figure 4.31: A cutaway of the McDonnell Douglas X-30 hypersonic aircraft shows the liquid hydrogen fuel storage and surface cooling system.
(Courtesy of McDonnell Douglas Corporation)

Figure 4.32: Fuel selection impacts size, weight & drag.

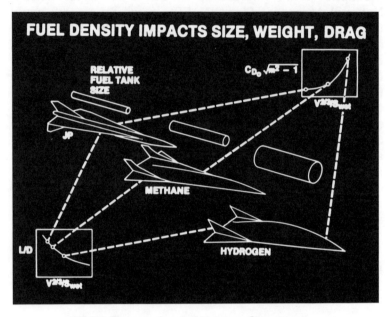

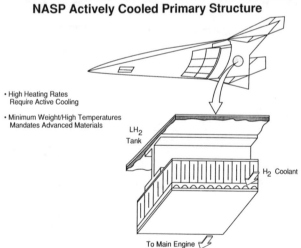

Figure 4.33: Liquid hydrogen cooling.
(Courtesy of McDonnell Douglas Corporation)

Whereas the Space Shuttle relies on expensive ceramic tiles to keep the surface of the spacecraft from melting during reentry, as Figure 4.33 indicates, the hypersonic aerospacecraft will use the ultra-cold liquid hydrogen fuel to cool its outer surface. While hypersonic aerospace vehicles offer a glimpse of what advanced hydrogen energy technologies will look like, the ultimate hydrogen energy transportation -- and living -- technology will probably be hydrogen-fueled space habitats.

*Space Habitats*

Even with a successful energy transition to renewable resources, it is only reasonable to assume that the other available resources on the Earth are limited; and as the human population continues to exponentially increase, it will become increasingly necessary to utilize the vast resources that exist in space. Dr. Gerard K. O'Neill, a professor of physics at Princeton University, has pointed out that the resources that abound in space are virtually inexhaustible. In his paper, "The Colonization of Space" published in *Physics Today*[30], and book, *The High Frontier: Human Colonies in Space*[31], he calculated that it would be possible to have more humans living in large space colonies than on the Earth by the year 2075. O'Neill and his colleagues did detailed studies on the construction of large-scale space habitats that would rotate to simulate Earth's gravity. This means large cities in space would allow humans and other mammals to remain in space for extended periods of time without having adverse physiological reactions. Without simulating the gravity of Earth, for example, the human body would soon turn into a shapeless, jellyfish type of organism.

One of the most far-sighted hydrogen-fueled spacecraft configurations has been conceived by R.W. Bussard, a member of the British Interplanetary Society. The Bussard spacecraft pictured in Figure 4.34 is called a interstellar ramjet "hydrogen scoop." This is because it is designed to scoop up free hydrogen atoms that drift in space between the stars and then accelerate them into fusion engines for propulsion. Such a configuration represents a renewable-fueled hydrogen spacecraft that could accelerate to nearly the speed of light.

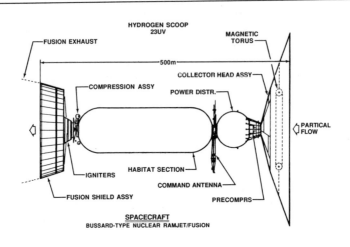

Figure 4.34: Blueprints for a Bussard "Hydrogen-Scoop" Ramjet Spacecraft.
(Illustration prepared by Rick Sternback [32])

It has been estimated by Carl Sagan in his book *Cosmos*, that in deep space, there is only about one hydrogen atom per ten cubic centimeters, which is a volume about the size of a grape. This means that if a hydrogen-fueled Bussard ramjet is to be able to work, it would need a frontal scoop that is perhaps 500 kilometers (i.e., 310 miles) across, and the engines would have to be the size of small cities [33].

If O'Neill's calculations are correct, building large structures in space is a straight-forward engineering problem. Building fusion-powered engines capable of accelerating a spacecraft to nearly the speed of light is, however, another matter. Even if the daunting technical obstacles of achieving such speeds were to be resolved, Sagan points out that other serious problems would be created by traveling at such high speeds. For example, hydrogen and other atoms would then be impacting the spacecraft at close to the speed of light, and such induced cosmic radiation could be fatal to any human passengers on board.  On the other hand, if a hydrogen-scoop space-craft were designed to use conventional hydrogen-oxygen fueled engines, in contrast to advanced fusion engines, it would probably not be able to achieve such high speeds, but it could be built with existing technologies.

It is worth noting that the primary reason for putting large numbers of people into space is to resolve the fun-damental global problems of more and more people com-peting for fewer and fewer resources.  Traveling at near the speed of light in order to reach distant galaxies is at best a secondary consideration.  This is especially true when one considers that by the time such advanced technologies be-come available, biotechnology will probably have provided life spans that are virtually indefinite, which means the need to hurry along will no longer be such an important consideration.  What follows is a general idea of how a state-of-the-art hydrogen-fueled biohabitat spacecraft might be designed.

### Starship Hydrogen

If it is necessary to have a hydrogen scoop that is at least 300 miles in diameter, the first consideration is that the scoop is such a substantial structure in terms of cost and materials that it would probably be modified to be-come part of a large biohabitat area pictured in Figure 4.36.  The whole concept of a "scoop" would probably not be necessary in any case.  The free-floating hydrogen atoms would probably be picked up by a thin membrane that would coat the outer skin of the frontal surface of the biohabitat area.  It would collect hydrogen atoms much like a photovoltaic solar cell collects photons from the Sun, and then shuttle them to a central receiver area.  Since oxygen is relatively rare in free space, large oxygen storage tanks would be necessary (refer to Figure 4.35).

The hydrogen and oxygen collected by such a space-craft would primarily be used for external propulsion and maneuvering.  This is because all of the hydrogen-fueled vehicles and megastructure cities inside the biohabitat area would be continually recycling the hydrogen used for energy purposes. Such a spacecraft would not be dependent on either solar or nuclear energy, and as such, it could travel through space indefinitely, scooping up hydrogen and oxygen like a primitive microbe in the primordial sea. Earth-like gravity (i.e., 1G) could be maintained on the spacecraft by having the biohabitat cylinder continually rotate. If it were necessary for 1G to be maintained on the smaller cylinder areas, they would need to be rotating at different speeds.

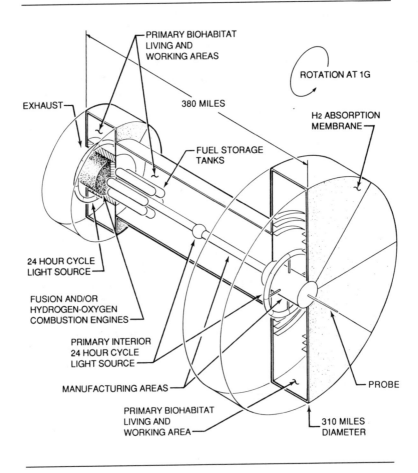

Figure 4.35: Starship Hydrogen Core Structure.

Figure 4.36: A prospective view from inside the starship.

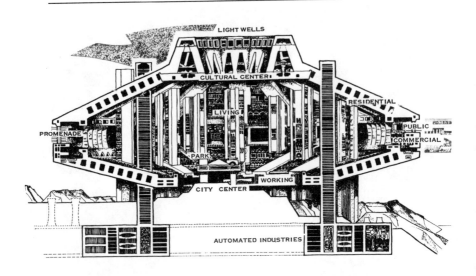

## Arcology
**by Architect Paolo Soleri**
Scottsdale, Arizona

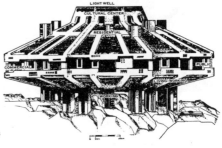

| Population | |
|---|---|
| Density | 340,000 |
| Height | 1,396/hectare; 565/acre |
| Diameter | 850 meters |
| Surface covered | 1,750 meters |
| Section and elevation | 240.8 hectares; 595 acres |
| Scale | 1:10,000 |

Figure 4.37: Biohabitat Arcology.
(Illustration reprinted with permission from Massachusetts
Institute of Technology at Cambridge)

The illustration in Figure 4.37 is a megastructure designed by architect Paolo Soleri (Scottsdale, Arizona). Soleri refers to such megastructures as "arcologies" (i.e., an
integration of architecture and ecology), and they provide a
perspective of what cities could be like both on the Earth,
and in a space biohabitat, assuming there would be few
automobiles. The urban area greatly shrinks in size, because without automobiles, there are no roads or gas sta-

tions on every corner. In fact, without automobiles, the whole character of the urban center changes dramatically. Residential areas would be located on the outer surfaces of miles-high arcologies, while shopping, working and other indoor areas would be housed deep inside the structure.

When it became necessary to travel to another arcology located in the distance, the principal vehicle would likely be a Vertical Take-Off and Landing (VTOL) aircraft like the Moller aircraft pictured in Figure 4.38 rather than automobiles. The Moller aircraft is not a science fiction dream. It is a state-of-the-art aircraft that is so different from conventional helicopters and fixed wing aircraft that it has a new name: *Aerobot* (which refers to the remotely controlled robotic capabilities). Various prototypes and designs have been developed over the past twenty years by Dr. Paul Moller, a former professor of aeronautical engineering at the University of California at Davis. Moller now has his own firm, Moller International, which manufactures various automotive components, including the internal-combustion rotary engines that are at the heart of the Moller aircraft.

Figure 4.38: Moller Vertical Take-Off and Landing Aircraft.
(Photo courtesy of Moller International)

A 2 or 4-seater Moller aircraft could be equipped with from three to six rotary engines that would be linked to an on-board computer so their temperature and revolutions per minute (rpm) could be continuously monitored. If all the engines should fail, the aircraft is to be equipped with a ballistically deployed parachute system, a shock-absorbing nose and fuselage, and, presumably, airbags. At a cruising speed of about 300 mph, the Moller aircraft will be expected to average about 15 miles per gallon, which would provide it with a range of about 750 miles on its 50-gallon fuel tank, yet it is small enough to fit into a standard garage. The rotary engines are quieter than conventional aircraft engines. When hovering at 50 feet, a Moller aircraft will be expected to emit about 85 decibels of sound, which is less than one-third that of a Cessna 150 during takeoff. The rotary engines can burn virtually any fuel, including hydrogen, and when Moller-type aircraft are mass-produced, they will be expected to cost no more than existing automobiles.

Although having so many vehicles in the air could result in numerous collisions, it is anticipated that a highly reliable computer collision-avoidance system will eventually be a part of every aircraft. In addition, most urban traffic density is a result of millions of people driving to work on relatively narrow roadways that allow only a few lanes of traffic to pass at any given time. It is worth noting that if urban dwellers lived in Paolo Soleri-type arcologies instead of the unplanned chaos that exists now in most cities, people would not need to use their vehicles in order to get to work, to school, or to go shopping. Although such a technological-utopia is surely possible in a futuristic hydrogen-fueled biohabitat space colony, it is also possible to have the same technological utopia here on Earth right now. It is not a question of technology, but rather political and social priorities in terms of determining what technologies should be developed.

As the human and animal populations continue to increase, it will become increasingly necessary to have people living in space. As Buckminster Fuller noted, humans going into space is somewhat like a chicken coming out of its egg. This being the case, it is likely that the inhabitants of such biohabitat colonies will want vast wilderness areas

to simulate the Earth as much as possible. The initial
Starship Hydrogen pictured in Figure 4.36, which would
be expected to have roughly the same land area as California and Arizona, could eventually become a small core of a
much larger (6000-mile diameter) biohabitat area pictured
in Figure 4.39. Such a structure would have ten times the
land area of the continental U.S.

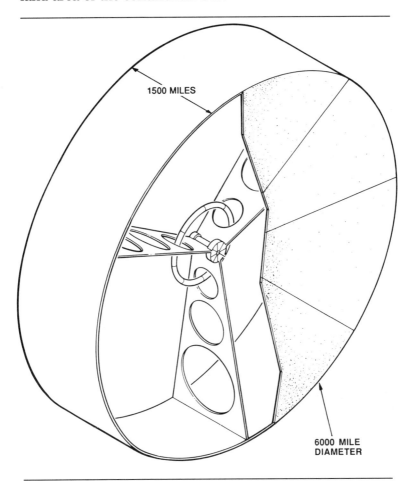

Figure 4.39: The Evolution of Starship Hydrogen.
Note the initial starship is now the hub of a much larger
biohabitat area of about 28 million square miles, which is
roughly 10 times larger than the continental U.S.

*Other Alternative Fuels*

There are many other alternative transportation fuels that could be used other than hydrogen. These include ethanol, methanol, natural gas (methane), ammonia or acetylene. When a switch to alternative or synthetic fuels is contemplated, it is especially important to examine all of the most viable options for two basic reasons:

1. Existing automotive and aerospace technology could be directed and optimized to utilize any of the alternative fuel options.

2. The cost of making an energy and industrial transition to any of the alternative fuels will be substantial.

The primary reasons for selecting hydrogen instead of one of the many other alternative fuel options is that hydrogen is the most environmentally acceptable fuel; it is compatible with virtually all other energy sources, and it is completely renewable. Although the minimal environmental impact of burning hydrogen in the atmosphere is one of its most obvious advantages, the fact that it is renewable is equally important, given the projected exponential increases in global energy consumption that were discussed in Chapters 2 and 3. In addition, hydrogen is one of the most efficient methods of storing and utilizing the vast but intermittent solar energy resources. Equally important is the fact that once a transition to a hydrogen energy system is made, there will never be a need to change to anything else. Dr. Howard Harrenstien, Dean of the School of Engineering & Environmental Design at the University of Miami, underscores the importance of hydrogen with the following observations:

> *"Many of the proposed new sources of primary energy, such as wind energy, ocean waves, ocean currents, ocean thermal differences, solar energy, and geothermal energy all suffer location disadvantages. They are generally not present in the*

*right magnitude at the right place and at the right
time for them to be directly converted to electricity
and placed on line at that point. Even if they could
be so converted, line losses imposed by transmission distances would place a heavy tax on efficiency.*

*Enter hydrogen. Ocean thermal electric plants,
which are located in the tropical seas, could
generate it on location and store it at hydrostatic
pressure on the sea bottom. Ocean tankers could
collect the product at the proper time and deliver it
to the world marketplace at great efficiencies in
energy efficiency. Floating stable platforms, using
photovoltaics and electrolysis to convert solar energy to hydrogen in mid-ocean, could also be
placed 'on line.' Even bottom of the sea geothermal
energy plants could be developed to produce hydrogen for these purposes...*

*In sum, the hydrogen economy may hold the
key to the integration of many new sources of energy into a common, environmentally acceptable
synthetic fuel -- one which will allow us to con-
serve our precious fossil fuel reserves and, at the
same time, develop a higher-level technology to
advance the quality of life in this country and the
world."* [34]

*Hydrogen Production*

Because hydrogen is the lightest of the 92 naturally
occurring elements, if it is not chemically bonded to heavier atoms, such as carbon or oxygen, it will likely drift up
into the upper atmosphere where it will eventually react
with oxygen to form water. It is for this reason that free
hydrogen does not exist in significant quantities within the
Earth's atmosphere. As a result, if hydrogen is to be used
as a fuel or chemical feedstock, energy must be expended
to separate the hydrogen atoms from the heavier atoms of
carbon or oxygen.

As stated earlier in this chapter, most of the hydrogen
currently used in the chemical and petroleum industry is

manufactured from natural gas, which is a hydrocarbon molecule with four hydrogen atoms bonded to one carbon atom. (Gasoline is also a hydrocarbon molecule that is made up of eighteen hydrogen atoms that are attached to a chain of eight carbon atoms.) High temperature steam is used to separate the hydrogen from the carbon, and if the cost of the natural gas is $2 per million British thermal units (mmBtu), the cost of the gaseous hydrogen will be about $5.00 per mmBtu[35]. If the hydrogen were to be liquefied on a large scale, an additional $4.00 to $5.00 per mmBtu would need to be added to the cost of the gaseous hydrogen[36], making the cost of liquid hydrogen produced by this method about $10.00/mmBtu. This corresponds to gasoline costing about $1.20 per gallon. If hydrogen is manufactured from water with off-the-shelf electrolysis equipment, its cost is roughly equivalent to $3.65/mmBtu per 10 mills (i.e., $3.65/mmBtu/cent/kWh).

Hydrogen can also be manufactured from coal-gasification facilities at a cost ranging from $7.85 to $11.20 per mmBtu, depending on the cost of coal and the method used to gasify it. But making hydrogen from nonrenewable fossil fuels does nothing to solve the basic problems of the exponentially diminishing resources, nor the serious environmental problems that will result from using such resources. Moreover, because the fossil fuels are nonrenewable, ever growing amounts of energy will be required to extract what is left. The easy-to-get oil has already been found, and increasingly, exploration efforts will have to drill deeper and deeper, and more and more dry holes will result. At some point, it will take more energy to extract the remaining fossil fuels than they will contain in their chemical bonds.

Given these realities, it is only realistic to think in terms of producing hydrogen from resources that are renewable, such as the direct and indirect sources of solar energy, which includes the large quantities of agricultural wastes, sewage, paper and other biomass materials that are accumulating in landfills. This is an important consideration because virtually all of the existing landfill sites in the U.S. will be filled to capacity before the year 2000. Moreover, generating hydrogen from such "waste" materials may turn out to be one of the least expensive methods of producing hydrogen -- and the resource is quite substantial. It has been estimated that in the U.S., roughly 14 quads of the annual 64 quad total energy requirement could be met from renewable biomass sources, which is roughly 20 percent of the total.

With respect to sewage, it is both tragic and ironic that the vast quantities (literally billions of gallons per day) of human and animal waste that is being dumped into rivers and the oceans could -- *and should* -- be recycled to produce a renewable source of hydrogen. Researchers have shown that this can be accomplished either by utilizing the non-photosynthetic bacteria that live in the digestive tracts and wastes of humans and other animals, or by pyrolysis-gasification methods[37, 38].

This is an especially important consideration because marine biologists are already warning that the microbial food-chains in the oceans are in serious trouble and may be about ready to collapse because of the enormous levels of sewage and other toxic wastes that are being dumped daily. Since roughly 80 percent of the oxygen in the Earth's atmosphere comes from the microbes that live in the oceans, having them suddenly disappear could be catastrophic for most oxygen-breathing organisms. This problem is compounded by the fact that most of the oxygen-producing microbes live primarily on the continental shelves where there is sufficient light for the microbes to run their photosynthetic processes. Unfortunately, these are the very areas where most of the raw sewage and other toxic chemicals are being dumped.

As such, the national security interests of the nation are now dependent on solving the human, animal and toxic waste problems. It should logically follow that instead of spending billions of dollars on B2 bombers and MX missiles, that money could and should be spent to develop advanced sewage treatment systems that could turn the billion of gallons of raw sewage that is being dumped into the oceans into relatively low-cost hydrogen.

Although high-temperature nuclear-fusion reactors may some day be practical as renewable sources of energy for hydrogen production, they are at present only a theoretical option. Typically, over 100 million degree Fahrenheit temperatures are required for nuclear fusion to occur, and as a result, such highly sophisticated technologies are not expected to be commercially viable for many decades, if not centuries. Before energy technologies can be considered economically viable, they must first be technically viable, and high-temperature fusion reactor systems are a long way from being either. Such energy options cannot be realistically expected to produce any significant quantities of hydrogen or electricity before the existing 10 to 15-year supply of known U.S. oil reserves are exhausted. There is always the possibility that new and innovative nuclear energy systems could be developed for safe and economical energy production purposes, but long-term energy decisions should not be predicated on nonexistent or unproven technologies.

*Conclusions*

An industrial transition from a petroleum economy to a hydrogen economy would resolve many of the most serious global environmental problems. The serious problems associated with fossil fuel and nuclear fission energy systems underscore the importance of developing renewable energy options that do not pose such long-term and unknown environmental and economic risks. Moreover, when one undertakes a careful review of the most viable renewable energy options that could generate enough hydrogen to displace the use of fossil and nuclear fuels, one is inevitably reduced to the relatively simple technology options that are in one way or another associated with solar or biological energy processes and resources. It is worth noting that biological organisms have been successfully utilizing a hydrogen energy system on a global scale for over 3.5 billion years.

Because solar-hydrogen technologies can be mass-produced by automotive industries, they should be able to produce energy that is economically competitive with that generated from fossil or nuclear energy systems, while providing significant private-sector employment and economic stability for many decades. Most importantly, such an energy transition will allow civilization's millions of industrial machines to function in relative harmony with the

Earth's natural biological life support systems. For-tunately, there are many viable solar options, which will be discussed in Chapter 5, that can be developed for large-scale hydrogen production. Because such systems are no more complex to manufacture than automobiles or ships, they can be rapidly implemented on a global-scale with existing technology. Willis M. Hawkins, a Senior Technical Advisor and former Board Member of the Lockheed Aircraft Corporation, provides a note of encouragement when he writes:

> *"I certainly hope that those who recognize the advantages of hydrogen for industrial and home consumption will be pressing just as hard on the infrastructure serving those customers as we will be for aircraft. We hope that the pioneers and developers in other fields will challenge us or even help us to lead in converting the world to hydrogen."* [39]

For a more detailed historical discussion of the hydrogen energy option, read *The Forever Fuel: The Story of Hydrogen* (Westview Press, Boulder, Colorado), written by Peter Hoffmann, a former deputy bureau chief of McGraw-Hill World News in Bonn, Germany. Hoffmann's book provides an excellent overview of the discovery and use of hydrogen energy systems. However, the most extensive single source of detailed technical information on the hydrogen energy option is provided in the *International Journal of Hydrogen Energy*, the official publication of the International Association for Hydrogen Energy (IAHE)[40].

The IAHE is a professional peer-review society involving some of the world's most distinguished scientists and engineers from within industry, education, and government. It is a technical brain-trust that has representatives from over 80 countries who have been carefully assembling the necessary engineering and chemical information that will be required when the world makes the decision to make an industrial transition to renewable resources a reality. Dr. T. Nejat Veziroglu, (University of Miami, Coral Gables, Florida), is president of the IAHE. He is also Editor in Chief of the IAHE Journal, and is also one of its original founding fathers. He has written that:

*"We do not have to subject the biosphere and life on this planet to the deadly effects of fossil fuels. The answer is to establish the environmentally most compatible, clean and renewable energy system, the Hydrogen Energy System...*

*What we need now is the governmental decisions to convert to the new energy system in an expeditious and prudent manner. When the subject is brought up with government officials, their answer is 'Hydrogen sounds good, but just now it is more expensive than petroleum. Let the free market forces decide what the energy system will be.' The answer is fine, if the rules of competition are fair. At the present time, they are not. The rules favor the lower production cost products, irrespective of their environmental effects. We need new and fair laws which take into account environmental damage, as well as production costs...*

*Once such a principle is legislated, then no energy company will produce and sell petroleum for fuel, but will begin to manufacture hydrogen, since it will be by far the cheapest fuel."* [41]

Dr. Carl Sagan is quoted in the beginning of this chapter as saying that we are all essentially made up of "starstuff." It seems appropriate, therefore, to conclude this cosmic discussion on hydrogen by expanding on Sagan's explanation of the origin and evolution of hydrogen in the universe. The following passage is quoted from Sagan's *Cosmos* television series that was aired on the Public Broadcasting System (PBS):

*"Some 15 billion years ago, our universe began with the mightiest explosion of all time. The universe expanded, cooled and darkened. Energy condensed into matter, mostly hydrogen atoms, and these atoms accumulated into vast clouds rushing away from each other, that would one day become the galaxies. Within these galaxies, the first generation of stars was born, kindling the energy hidden in matter, flooding the Cosmos with light. Hydrogen atoms had made suns and starlight.*

*There were in those times no planets to receive the light and no living creatures to admire the radiance of the heavens. But deep in the stellar furnaces, nuclear fusion was creating the heavier atoms; carbon and oxygen, silicon, and iron. These elements, the ash left by hydrogen, were the raw materials from which planets and life would later arise. At first, the heavier elements were trapped in the hearts of the stars, but massive stars soon exhausted their fuel, and in their death-throes returned most of their substance back into space. Interstellar gas became enriched with heavy elements. In the Milky Way galaxy, the matter of the Cosmos was recycled into new generations of stars, now rich in heavy atoms, a legacy from their stellar ancestors. And in the cold of interstellar space, great turbulent clouds were gathered by gravity and stirred by starlight. In their depths, the heavy atoms condensed into grains of rocky dust and ice, complex carbon-based molecules. In accordance with the laws of physics and chemistry, hydrogen atoms had brought forth the stuff of life.*

*Collectives of organic molecules evolved into one-celled organisms. These produced multi-celled colonies, when various parts became specialized organs. Some colonies attached themselves to the sea floor. Others swam freely. Eyes evolved, and now the Cosmos could see. Living things moved on to colonize the land. Reptiles held sway for a time, but they gave way to small warm-blooded creatures with bigger brains who developed a dexterity and curiosity about their environment. They learned to use tools and fire and language. Starstuff, the ash of stellar alchemy, had emerged into consciousness...*

*These are some of the things that hydrogen atoms do, given 15 billion years of cosmic evolution. It has the sound of epic myth, but it's simply a description of the evolution of the Cosmos, as revealed by science in our time."* [42]

Figure 4.40: Sun & Sea/Solar Hydrogen.

Chapter 5

# SOLAR TECHNOLOGIES

There are many direct and indirect solar technologies that have been under investigation by energy engineers for the past several decades, and in many cases, specific technologies have been developed for energy production and testing. A quick review of some of the current solar options that could produce base-load quantities of electricity and/or hydrogen include the following:

* Photovoltaic cells, which are able to convert sunlight directly into electricity with no mechanical moving parts.

* Wind energy systems are indirect solar technologies that can be deployed both on land and out at sea.

* Ocean thermal energy conversion systems (OTEC) are indirect solar systems that utilize the vast amounts of solar heat that is stored near the surface of the oceans.

* Ocean currents, tides and waves are also indirect forms of solar energy that could be utilized, although the technology to do so is not very well developed.

* Solar thermal systems, which utilize the Sun's heat energy. Solar thermal examples include line-focus "trough" or point-focus-concentrator "dish" and power tower systems.

In evaluating the various solar options, or any energy technology for that matter, the primary concern is economics. There are many people who philosophically support the use of solar energy, but if the energy produced from solar technologies is three or four times more expensive than energy produced from the highly polluting fossil or nuclear fuels, most of those same people will use the latter and not the former.

It is generally assumed that solar energy technologies are not presently economically competitive with conventional fossil fuel or nuclear facilities, but this assumption is both inappropriate and inaccurate. The assumption is inappropriate because fossil fuels are non-renewable, which means their prices are inherently unstable. While the surplus of oil reduced its cost for the decade of the 1980's, it is only a question of time before the gas lines and high prices return. It is well to remember that the time to fix the roof is when it has stopped raining. To use the current price of fossil fuels in terms of long-range energy policy formation is like trying to drive a car while continuing to look in the rear view mirror. One does not realize one is in trouble until after an accident occurs.

The assumption is inaccurate because the environmental costs of using nuclear and fossil fuels are not factored into the energy cost equation. This is not only because the environmental costs are staggering, but in many cases, they are difficult, if not impossible, to calculate. However, even without considering the environmental costs, there are many solar technologies that will be discussed in this chapter that could be developed, and could produce electricity and hydrogen at prices that are competitive with energy produced by both fossil-fueled and nuclear facilities.

Unfortunately, most members of the U.S. Congress, and the public in general, are unaware of the most cost-effective solar technologies that could fundamentally resolve many of the most serious economic and environmental problems threatening to make the Earth uninhabitable. Indeed, the only solar technology that has received widespread exposure and most of the U.S. government research dollars, happens to be one of the least cost-effective: photovoltaic cells.

*Photovoltaic Cells*

Photovoltaic solar cells are semiconductor devices that are able to convert sunlight directly into electricity with no moving parts. Because photovoltaic cells have successfully provided electricity for space vehicles for many decades, as well as a wide range of consumer electronic devices such as calculators and watches, they have received most of the attention in the media. Indeed, for most people, when solar-electric energy systems are mentioned, they assume one is referring to photovoltaic systems. The high costs involved in manufacturing photovoltaic systems, however, have thus far prohibited their use for large-scale energy production purposes.

Figure 5.1: A photovoltaic solar cell array provides electricity with no moving parts for a residence in the desert Southwest.

The "photoelectric effect" was first observed in 1881 by the German physicist, Heinrich Hertz. He discovered that light (which consists of the visible wavelengths of electromagnetic radiation) could displace electrons from certain metals. A photovoltaic solar cell is typically made of silicon, which acts as both an electrical insulator and a conductor. The silicon strips are coated with other materials, such as boron, to produce a positive electrical layer which interacts with the underlying negatively charged silicon layer. This positive-negative area is referred to as the "*p-n junction*," and when a photon of sufficient energy impacts an electron in the silicon, the electron moves across the p-n junction, and in doing so, produces a flow of direct-current (DC) electricity.

Figure 5.2: A large centralized photovoltaic array.
Such systems can easily be created by connecting smaller modularized panels to each other.

Although the cost of photovoltaic cells have been reduced from roughly $50 per peak watt in 1974, to less than $5, they are still too expensive for large-scale energy production. Electricity generated from fossil fuel and nuclear plants generally costs around $0.50 to $0.70 per peak watt, or roughly $0.06 to $0.08 per kilowhatt hour (kWh). In addition, photovoltaic cells must be attached to a support structure and other components that make up a complete system, which means their actual installed costs are presently in the range of $8.00 to $10.00 per peak watt, or roughly $0.25 to $0.30 per kWh. If hydrogen were generated with electricity that expensive, the cost would be comparable to gasoline costing roughly 10 to 12 dollars per gallon. However, most investigators in photovoltaic technology believe that the cost of photovoltaic systems will continue to be reduced, and that within another decade, photovoltaic systems will likely be competitive with conventional fossil fuel and nuclear power plants.

Typical silicon-based photovoltaic cell efficiencies have been in the range of 10 to 12 percent, although photovoltaic researchers at the Boeing Company have been able to achieve efficiencies of up to 37 percent by utilizing gallium arsenide instead of silicon. This is in contrast to nuclear or fossil fuel power plants that typically operate with a conversion efficiency of about 38 percent. On the other hand, nuclear and fossil fuel facilities waste 62 percent of the expensive and non-renewable fuels they require, and they leave behind large amounts of toxic waste products.

Although a photovoltaic cell may only be 10 percent efficient in converting sunlight into electricity, it is not wasting any non-renewable resources or producing any toxic by-products in the process. Thus, to compare solar efficiencies with the efficiencies of conventional power plants is inappropriate. What is appropriate, however, is to compare the efficiencies of other solar technologies that have been developed in recent years, because while photovoltaics have received the the vast majority of solar research funds, they are still one of the most expensive solar technology options available.

While there will eventually be an efficient -- and more importantly, a low-cost photovoltaic system developed -- if existing photovoltaics were the only solar option available, the industrial transition to renewable resources would not be economically viable. Fortunately, there are economical solar technology options other than photovoltaic systems. Three of the most promising solar technologies include wind energy conversion systems, ocean thermal energy conversion (OTEC) systems, and point-focus-concentrator "dish" systems.

*Wind Energy Conversion Systems*

Wind machines were the first solar technology to be used for mechanical power. One of the first uses of wind machines occurred around the year 650 A.D. when Persian millwrights figured out how to use the wind to pump water for irrigation. When Genghis Khan's nomadic warriors swept through Persia in the 1200's, they captured many of the millwrights and the windmills they had constructed, and took them back to China. To this day, similar windmills are still used to pump water. Windmills were first used to pump water in Europe in the early 1100's, and they came into widespread use in the U.S. for water pumping during the early 1800's.

With the development of electrical generating equipment in the late 1800's, both European and American engineers began to experiment with using the wind to operate electrical generating equipment. It was not long before windmills evolved into wind generators. One of the first investigators to develop wind-powered electrical generators was a Danish professor, Poul La Cour, who experimented with wind systems from 1891 until his death in 1908. He was one of the first individuals who foresaw the use of hydrogen as a fuel and the use of wind-powered electrical generators to electrolyze hydrogen and oxygen from water. Another early investigator who promoted wind-powered hydrogen production systems was J.B.S. Haldane. Haldane was a British biochemist at Cambridge, England, who, in 1923, predicted that England's future energy problems would be solved by constructing large numbers of metallic wind generators that could supply high-voltage electricity to large electrical mains for hydrogen production.

During World War II, Vannovar Bush, a physicist who was the director of the U.S. Wartime Office of Scientific Research and Development, became worried over dwindling American fuel reserves and concluded that wind generators might be an answer. He appointed Percy H. Thomas, a wind power advocate, to the Federal Power Commission, which later convinced the Department of the Interior to construct a large prototype wind machine. However, in 1951, the idea died in the House Committee on Interior and Insular Affairs. Wind-generated electricity could not compete economically with coal that was selling for $2.50 per ton or diesel fuel that was $0.10 per gallon. The promise of even less expensive electricity ("too cheap to meter") from nuclear power plants resulted in the abandonment of virtually all Federal programs to develop wind-powered energy systems[1].

In retrospect, this decision turned out to be a tragic mistake. Although nuclear plants were initially predicted to cost between $250 and $300 per installed kilowatt (kW), in contrast to wind energy conversion systems that were expected to cost $400 to $500 per installed kilowatt (kW), the nuclear plants ended up costing $1,500 to $2,500 per installed kW. Moreover, the capital costs for nuclear power plants did not include the extensive "front-end" costs of uranium enrichment, nor the "back-end" costs associated with decommissioning the plant at the end of its useful life, or the long-term waste storage problem.

In contrast, a NASA study completed in 1972 found that wind machines are a historically mature technology that is based on the straightforward application of well-understood principles and practices in areas of civil, electrical, structural, corrosion and aerodynamic engineering. As a result, any manufacturing problems likely to arise can be subjected to a body of existing engineering knowledge and experience, and thereby rapidly resolved.

The validity of this statement is supported by the fact that a 1.2-megawatt Smith-Putnam wind generator, built by the Central Vermont Public Service Company in 1940, took only two years from the drawing board to power production. This contrasts with nuclear-fission power plants that usually take more than a decade to design and build. In addition, wind-energy conversion systems can be mass-produced using airplane technology, and major aerospace firms such as Boeing, Grumman, Lockheed and General Electric are willing and able to manufacture large wind generators for utilities. This underscores the point that any future electrical demand could easily be met by wind-powered electrical generation systems.

*Wind Power Disadvantages*

Although wind power systems have many significant advantages, land-based wind machines have two major disadvantages. The first major drawback is that the topographic features on land can dramatically reduce wind speed, which significantly reduces the electrical output of a wind machine. This is because wind-powered machines conform to the "*Law of the Cube.*" This mathematical formula provides that the power content of the wind is proportional to the cube of the wind speed. This means that when wind speed is reduced by 50 percent, the power output of a wind machine is reduced by eight times.

The second major drawback of traditional land-based wind generators is that the best sites to obtain maximum wind speeds are high on mountain ridges, and putting thousands, much less millions of wind generators in scenic mountainous areas is, understandably, not a pleasant thought. Radioactive wastes may last for tens of thousands of years, but they are invisible, and for many people, that makes them more acceptable (i.e., out of sight, out of mind). In any case, there are alternative wind power systems that avoid these two major disadvantages.

One option involves placing wind generators out at sea. Such a proposal has been developed by an American wind power engineer, William E. Heronemus, a professor and associate head of civil engineering at the University of Massachusetts. Heronemus graduated from the U.S. Naval Academy and worked in the largest ship design group in the U.S. until he retired from the Navy in 1965. However, Heronemus is perhaps best known for his design of an offshore wind power system that would eliminate the basic disadvantages of land-based systems.

The system proposed by Heronemus would consist of a vast network of wind generators on floating platforms or concrete-pile Texas towers as pictured in Figure 5.3. Offshore wind machines are substantially more efficient because as winds leave land masses, they show a marked ability to intensify in velocity due to the fact that there is essentially nothing to slow them down. The offshore wind power system would consist of 83 wind units, and each unit would be made up of 164 wind stations that would have three 600 to 2000 kilowatt wind generators. These generators would be arranged in concentric rings around a centralized electrolyzer station, and the hydrogen that would be produced from the seawater would then be piped ashore and used as fuel for transportation vehicles or the production of electricity in conventional power plants.

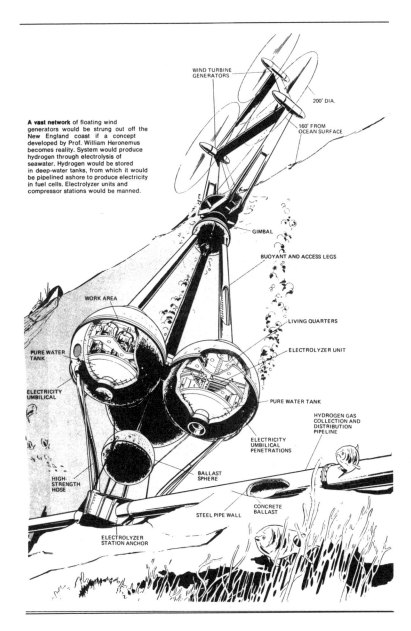

A **vast network** of floating wind generators would be strung out off the New England coast if a concept developed by Prof. William Heronemus becomes reality. System would produce hydrogen through electrolysis of seawater. Hydrogen would be stored in deep-water tanks, from which it would be pipelined ashore to produce electricity in fuel cells. Electrolyzer units and compressor stations would be manned.

WIND TURBINE GENERATORS

200' DIA.

160' FROM OCEAN SURFACE

GIMBAL

BUOYANT AND ACCESS LEGS

WORK AREA

LIVING QUARTERS

ELECTROLYZER UNIT

PURE WATER TANK

ELECTRICITY UMBILICAL

PURE WATER TANK

HYDROGEN GAS COLLECTION AND DISTRIBUTION PIPELINE

ELECTRICITY UMBILICAL PENETRATIONS

BALLAST SPHERE

HIGH-STRENGTH HOSE

CONCRETE BALLAST

STEEL PIPE WALL

ELECTROLYZER STATION ANCHOR

Figure 5.3: Offshore Wind Energy Systems designed by William E. Heronemus, Department of Engineering, University of Massachusetts at Amherst.

Studies have shown that the offshore wind-powered systems proposed by Heronemus could generate over 350 billion kilowatt-hours of electricity annually, which is about one-fifth of the electricity consumed in the U.S. The capital costs are estimated to be competitive with existing fossil fuel or nuclear systems, and the environmental impact of such systems would be minimal. Consider the typical environmental concerns of conventional power plants in Table 5.1 that would not exist with wind systems:

Table 5.1:
Environmental Emissions of Wind Machines.

| | |
|---|---|
| Thermal Pollution: | 0 |
| Sulfur Dioxide Emissions: | 0 |
| Nitrogen Oxide Emissions: | 0 |
| Carbon Monoxide Emissions: | 0 |
| Carbon Dioxide Emissions: | 0 |
| Hydrocarbon Emissions: | 0 |
| Radioactive Wastes Produced/Year: | 0 |
| Other Toxic Chemicals Produced/Year: | 0 |

Not only are wind power technologies essentially pollution-free in their operation, but so are most other renewable solar energy technologies that harness the forces of nature. Although offshore wind energy systems would involve some disturbance of the seabed during construction, no disturbances would occur during normal operations. Indeed, when one considers that the fisheries in the Earth's oceans are being wiped out by the over-fishing that is made easy by driftnetting practices and modern electronic fishing vessels, one realizes that large numbers of offshore wind power systems could provide a much needed sanctuary for the unprotected marine organisms. This is no small consideration.

A generation ago, New England fishermen were taking in thousands of tons of haddock and cod every year, and they were hardly denting the supply. But today, the haddock is gone from the New England coast, just as the herring has been hunted out of the vast North Sea. In what has been referred to as "stripmining the seas," thousands of fishing vessels are now devastating the once fertile fishing areas on a global scale. Even more ominous is the fact that more and more of these ships are being put into service every year. If large numbers of wind turbines were deployed at sea, they would help to offset the exponential destruction of the Earth's ocean ecosystems.

### Vertical Vortex Wind Generators

Another wind energy system that resolves the basic problems associated with traditional land-based systems has been developed by Dr. James T. Yen when he was working as a fluid dynamics research engineer at Grumman Aerospace Corporation (Whiteplains, New York). Yen called his innovative system a "*vertical vortex generator*," although it has been referred to as a "tornado turbine" because it operates on a similar principle.

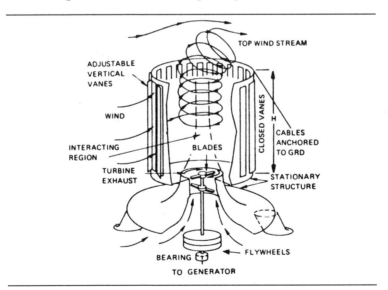

Figure 5.4: Vertical Vortex Wind Generator.
(Reprinted from *Popular Science* with permission © 1977 *Times Mirror Magazines, Inc.*)

Figure 5.5: An Urban Vortex Generator.
(Reprinted from *Popular Science* with permission © 1977
*Times Mirror Magazines, Inc.*)

The artist's conception in Figure 5.5 provides a view of what a vortex energy system might look like in a major metropolitan area. By placing vortex generators in urban areas, they can take advantage of the higher than normal winds that are often generated by large skyscrapers which channel the wind into narrow passage ways.

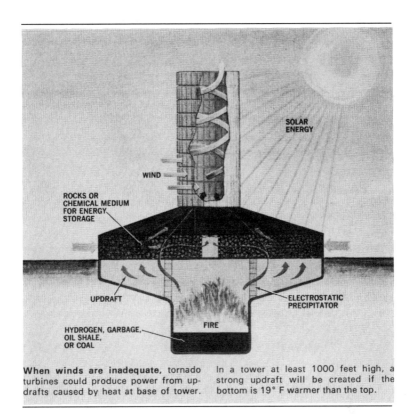

When winds are inadequate, tornado turbines could produce power from updrafts caused by heat at base of tower. In a tower at least 1000 feet high, a strong updraft will be created if the bottom is 19° F warmer than the top.

Figure 5.6: Vertical vortex generators can be powered by a wide-range of alternative fuels, including hydrogen. (Reprinted from *Popular Science* with permission © 1977 *Times Mirror Magazines, Inc.*)

Yen's system operates by constructing a large tower, and as wind enters the tower through louvers in its side, a vortex of air is created. A vortex is a swirling mass of air (or water) that forms a low pressure vacuum at its center, and anything that is caught in the motion is drawn in. This is how whirlpools and tornadoes operate, and the reason they develop such awesome power is because the wind pressure energy in a 15 mph wind is about 3,000 times greater than the kinetic energy that drives conventional wind energy systems. Thus, while a 200-foot diameter blade on a conventional wind turbine can produce approximately one megawatt, the same size blade on a vortex generator could produce from 100 to 1000 megawatts[2]. Because of the low pressure of the vortex, air is sucked in through the openings in the bottom of the tower, which then spins a turbine to generate electricity. Moreover, if there is insufficient wind, air heated by sunlight, hydrogen or even burning garbage can be used to operate the turbine. Although no large prototype vortex generators have been built, they offer a viable option that could be used in major metropolitan areas where conventional wind generators would be impractical. In this regard, architect Paolo Soleri has designed a megastructure city pictured in Figure 5.7 that could have a vortex generator integrated into the heart of the structure.

Figure 5.7: A Vortex Arcology. This megastructure was designed by architect Paolo Soleri. (The illustration was reprinted with permission from the Massachusetts Institute of Technology)

The idea that wind energy conversion systems could, and should, provide a major portion of the energy demanded by a modern industrial society has been well documented. Many competent engineers in many different countries have shown that the resource is vast, and the technical utilization is quite practical. As Heronemus has written:

> "In the oceanic winds, we have a huge energy resource that modern technology can harness to serve our needs on demand. It could be put to use in the very near term, economically, in an aesthetically satisfactory way, and with no pollution of any kind." [3]

### OTEC

Roughly forty-five percent of all solar radiation that reaches the surface of the Earth is absorbed by the surface water of the tropical oceans. Although the heat capacity of water is greater than that of any other fluid, the energy from the Sun cannot penetrate very deeply into ocean water. This is why cold water underlies even the warmest of the tropical seas. Ocean Thermal Energy Conversion (OTEC) systems not only can tap this enormous energy potential, but they are the only solar technology that can operate 24-hours a day, seven days a week, regardless of the time of year or weather conditions. This is because OTEC systems operate by taking advantage of the constant temperature differential that exists between the solar-heated surface water (which is about 80 degrees Fahrenheit) and the cold deep water (which is about 40 degrees Fahrenheit) to produce energy.

The Second Law of Thermodynamics states that the conversion of heat into mechanical work is possible when two heat reservoirs of different temperatures are at one's disposal. Hence the tropical oceans become a prime choice for thermal energy conversion systems. They have enormous heat capacity and a constant temperature differential of about 40 degrees that is permanently maintained by natural forces.

Figure 5.8: Ocean Thermal Energy Conversion (OTEC).
(Courtesy of Lockheed Missiles & Space Company, Inc.)

Capital costs for OTEC plants have been estimated to be anywhere from $1,000 to $4,000 per installed kW, depending on the design and number of units produced. Electricity costs are estimated to range from $0.04 to $0.10 per kWh. The mini-OTEC plant pictured in Figures 5.11 and 5.12 is a state-of-the-art prototype that was developed by Lockheed under contract with the U.S. Department of Energy during the Carter Administration. Although the tests proved successful, the program was terminated when the Reagan Administration concluded that the free market forces and not the Federal Government should determine national energy policy. The problem with that reasoning is that when Federal funding for OTEC research and development stopped, so did the research and development.

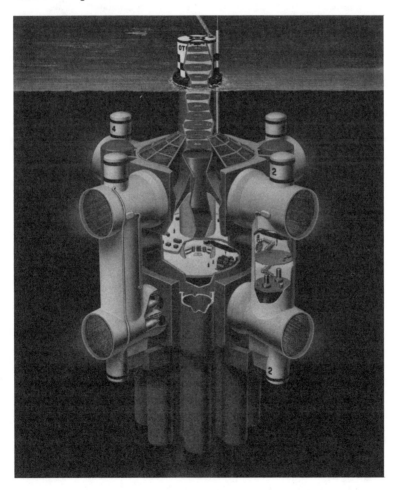

Figure 5.9: Lockheed OTEC: Cutaway View.
(Courtesy of Lockheed Missiles & Space Company, Inc.)

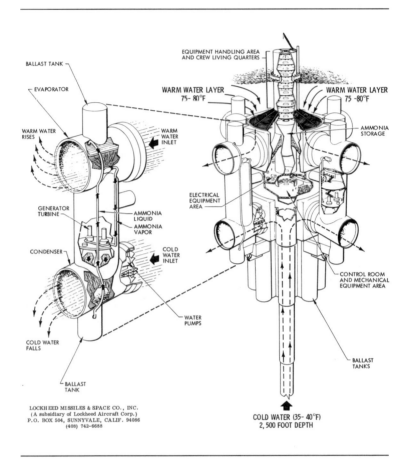

Figure 5.10: OTEC Operation.
(Courtesy of Lockheed Missiles & Space Company, Inc.)

Rather than developing renewable resource systems, the U.S. and the Soviet Union continued to spend billions of deficit financed dollars building ever more sophisticated weapon systems. In contrast, OTEC systems are relatively simple to build and maintain because they operate in low-temperature seawater environments. This means that any company that can build a supertanker or an offshore drilling rig can build an OTEC system.

Figure 5.11: A Prototype Mini OTEC Plant.

Figure 5.12: Mini OTEC Operation Primary Components.
(Courtesy of Lockheed Missiles & Space Company, Inc.)

With OTEC systems, the heat from the warm surface water is used to boil a working fluid, which creates a high-pressure vapor that is then used to spin a turbine. The turbine then drives a generator to produce electricity and/or hydrogen. The vapor is then chilled by the cold sea water that lies roughly 1,500 to 2,000 feet below the surface, until it turns back into a liquid, at which time it is then pumped back to repeat the cycle.

The same basic system is used in conventional coal or nuclear power plants; the only difference is the working fluid. Where conventional power plants use water as a working fluid to make high-temperature steam, most OTEC designs use a liquid that boils at or below the temperature of the surface sea water, such as ammonia, propane or one of the fluorocarbons. Since OTEC systems use a closed cycle, none of the working fluid ever escapes during normal operation [4]. Other significant advantages of OTEC systems include the following:

1. OTEC systems, in and of themselves, could produce enough energy to run the world (refer to Chapter 6 for details).

2. OTEC plants (or ships) can be built in large numbers in existing shipyards in a manner similar to the mass construction of the Liberty ships of World War II.

3. Sea-going systems do not require the purchase of real estate.

4. No high temperatures are involved in OTEC systems. Thus, less expensive materials can be used in their construction.

5. OTEC systems have been designed by numerous engineering firms, including Grumman Aerospace, Lockheed, Bechtel, TRW, and Sea Solar Power. Academic OTEC research teams exist at Carnegie-Mellon University, Johns Hopkins University, and the University of Massachusetts.

The OTEC concept is over 100-years old. It was first suggested as a source of power by the French physicist d'Arsonval in 1881, and the first OTEC pilot plant was constructed in 1930, at Mantanzas Bay, Cuba. Although the pilot plant was able to generate 22 kilowatts of power, the low cost of fossil fuels prevented any serious interest in developing OTEC power plants further. There is no question, however, regarding OTEC's vast energy potential. Moreover, as Dr. W.H. Avery, the Director of Ocean Energy Programs at the Johns Hopkins University Applied Physics Laboratory (Laurel, Maryland) has pointed out, OTEC plants are not dependent on any high technology breakthroughs. Rather, they essentially involve elaborate plumbing and low-temperature pumping systems[5].

In addition to their obvious purpose of electric power generation, OTEC systems have two other remarkable advantages. First, OTEC plants can desalinate sea water at a fraction of the normal cost. Investigators at Sea Solar Power (York, Pennsylvania), have calculated the cost of desalinated sea water at about four cents per thousand gallons. This means it would be less expensive for most major coastal cities to have fresh water barged in from OTEC plants rather than acquire it from their own potentially contaminated municipal water systems.

The second remarkable characteristic of OTEC power plants is they can greatly enhance fish yields. This is because the cold deep water that OTEC plants pump up to condense their working fluid back to a liquid, is rich in the nutrients that are necessary for aquatic plant and animal life. Indeed, natural cold water upwellings are responsible for some of the most fertile fishing grounds in the world, such as those off of the west coast of South America.

This means that deploying large numbers of OTEC plants throughout the tropical seas could dramatically increase world seafood supplies with vast open sea plant and fish farming areas[6,7]. According to researchers at Columbia University's Lamont-De-Herty Geological Observatory, the condenser effluent from a relatively small 100-megawatt OTEC plant could yield about 129,000 tons (wet weight) of shellfish meat annually, and a similar quantity of carrageen-containing seaweed. The current wholesale prices of shellfish meat exceed $2.00 per pound.

Thus, the value of the shellfish alone could exceed 500 million dollars, which is over 5 times the estimated $100 million capital cost of the OTEC plant itself.

Given all of the apparent advantages of OTEC systems, one can only wonder why it is nuclear technology that continues to receive the vast majority of energy research and development dollars. One thing *is* clear: between the winds in the atmosphere and the thermal differences that lie within the oceans, there is a vast resource of pollution-free energy. William Heronemus and his Amherst, Massachusetts engineering design teams have worked extensively on both OTEC and wind energy systems, and in summarizing his views on these renewable energy systems, Heronemus has written:

> *"These processes are the way of the future. When combined with other solar energy processes, they constitute the only energy regime which can sustain any real growth without making our planet uninhabitable."* [8]

Both wind energy and OTEC systems have the potential of generating electricity and hydrogen that could be competitive with fossil fuel or nuclear energy systems. Moreover, unlike fusion energy systems, neither of these renewable resource technologies require any major technological breakthroughs for their large-scale implementation. There is, however, a third solar technology that also has the potential to produce enough energy to displace fossil fuel and nuclear energy systems: point-focus-concentrator generator sets (gensets).

### Dish Genset Systems

Point-focus "dish" gensets, which look similar to a satellite or radar dish, are one of the most cost-effective solar options because they can be mass-produced with existing automotive and aerospace technology. Although dish gensets are in some respects like technological trees, in that they will use sunlight and water to make hydrogen, from a manufacturing and cost perspective, they are more like automobiles[9]. This is an extremely important consideration because the industrial expertise to manufacture automobiles is well-established.

Figure 5.13: Solar Dish Gensets "on Sun" at Edwards Air Force Base, California. They were developed in 1980 by NASA in cooperation with The Jet Propulsion Laboratory and The U.S. Department of Energy.

The solar dish system acts like a large magnifying glass which concentrates the incoming solar energy onto the power conversion units (i.e., engines) which can then generate electricity and/or hydrogen. It is worth noting that two of the engines used in the JPL tests pictured in Figure 5.13 were not initially developed for the prototype dish gensets, but for automobiles. The point-focus concentrator in Figure 5.14 is a more advanced design that was developed by McDonnell Douglas Corporation. Its Stirling engine power conversion unit, manufactured by United Stirling of Sweden, was also initially developed for automotive applications.

Figure 5.14: A state-of-the-art dish genset system.
A much larger point-focus concentrator system, referred to as a "power tower" is glowing in the background.
(Photo courtesy of Southern California Edison Company)

The individual sections of dish genset systems could be made of non-strategic materials, such as glass and steel, and they would be about the same size as the hoods and fenders for automobiles. Individual solar gensets can be connected together for large-scale electricity or hydrogen production, and unlike nuclear or fossil fuel systems, they would be able to produce electricity without any water requirements or the production of any toxic wastes.

Figure 5.15: Solar Dish Forest.
(Illustration provided courtesy of U.S. Naval Research Laboratory)

The illustration in Figure 5.15 is a painting by Pierre Mion that provides a prospective view of what a large field or "forest" of solar gensets sets might look like in a typical desert region. The solar gensets are like "technological trees" because they will be able to use sunlight to break water down into hydrogen and oxygen. Rather than a traditional glass and steel automobile-type dish concentrator, however, a low-cost concentrator that is built more like a bicycle (pictured in Figures 5.16, 5.17, and 5.18) may provide a more accurate picture of what the solar concentrators will look like in a future dish forest.

When the entire dish genset system is mass-produced by companies like Chrysler, General Motors, or Honda, a 20-kilowatt (kW) unit that could power a home and provide fuel for the family automobile would be expected to cost less than an average automobile. This is because of its simpler design, and the fact that it will not need to be changed for cosmetic reasons to enhance sales volumes. The fact that dish gensets are similar to automobiles from a manufacturing perspective means economic and engineering assumptions can be directly extrapolated from the experience of the automotive industries.

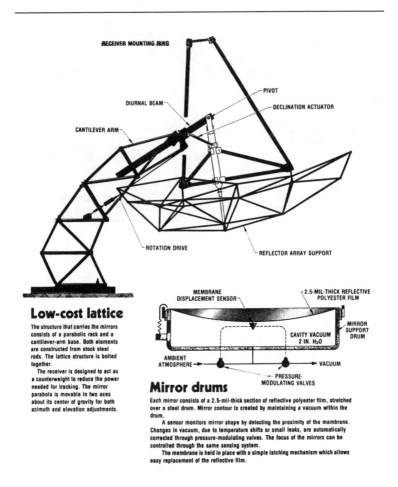

**Low-cost lattice**

The structure that carries the mirrors consists of a parabolic rack and a cantilever-arm base. Both elements are constructed from stock steel rods. The lattice structure is bolted together.

The receiver is designed to act as a counterweight to reduce the power needed for tracking. The mirror parabola is movable in two axes about its center of gravity for both azimuth and elevation adjustments.

**Mirror drums**

Each mirror consists of a 2.5-mil-thick section of reflective polyester film, stretched over a steel drum. Mirror contour is created by maintaining a vacuum within the drum.

A sensor monitors mirror shape by detecting the proximity of the membrane. Changes in vacuum, due to temperature shifts or small leaks, are automatically corrected through pressure-modulating valves. The focus of the mirrors can be controlled through the same sensing system.

The membrane is held in place with a simple latching mechanism which allows easy replacement of the reflective film.

Figure 5.16: LaJet dish Concentrator. This system utilizes low cost steel tubing and a reflective polyester film that is stressed over aluminum rims. A vacuum is used to maintain the film in a concave shape necessary for precise reflection. (Courtesy of *LaJet Energy Company*)

Figure 5.17: The LaJet Concentrator: Bicycle tubing and a nuts and bolts assembly. (Courtesy of *LaJet Energy Co.*)

Figure 5.18: SolarPlant 1.
700 LaJet dish concentrators generate high-temperature steam to power a conventional 5-megawatt electrical generating system in Warner Springs, California. (Reprinted from *Popular Science* with permission © 1985 *Times Mirror Magazine & Jeff Austin Photography* [10])

The fact that solar gensets will be able to produce electricity without any water requirements or the production of any toxic or radioactive wastes is reason enough to encourage their development. But in addition, dish Stirling gensets can be mass-produced in automotive and aerospace industries. Thus, the electricity they generate is expected to be substantially less expensive than the electricity generated by either fossil fuel or nuclear facilities. Calculations indicate that if 20 to 25 twenty-kilowatt solar units are installed per acre, roughly 10 percent of the desert areas in the American Southwest could make the U.S. energy independent of the Earth's remaining fossil fuel and uranium reserves.

### Early Dish Stirling Systems

The first known dish energy conversion units were designed and built in the U.S. right after the Civil War by the engineer and inventor, John Ericsson. Ericsson was fascinated by the potential of solar-powered engines, and by 1875, he had developed seven different types of what he referred to as "sun motors." Ericsson's first sun motors had incorporated steam engines for power production, but because of the continual valve failures he experienced, his later designs utilized closed-cycle Stirling engines. In a letter sent to a friend and business associate in 1873, Ericsson explained that in contrast to his earlier models, his new Stirling engine design was absolutely reliable. By 1880, he had developed his Stirling sun motor system into a conventionally-fueled water pump, and thousands of the machines were used worldwide.

The Stirling engine has a unique and colorful history of its own. It was invented in 1816 by the Reverend Robert Stirling, a Scottish minister who eventually became the Father of the Church of Scotland. Dr. Stirling was a noted classical scholar as well as an esteemed scientist, and he came from a family of distinguished engineers. His technical innovation was to reuse heat that would otherwise be wasted in conventional engines. What he created was one of the most efficient thermodynamic mechanical cycles ever developed: the Stirling cycle[11].

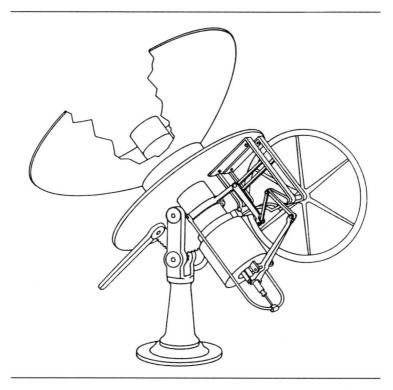

Figure 5.19: A drawing of a Stirling "sun motor" designed and built by John Ericsson in 1872 [12].

Conventional gasoline-fueled engines use an "Otto cycle" homogeneous-charge combustion process that was developed in 1870 by Nikolaus Otto. Diesel-fueled engines, on the other hand, utilize a more efficient stratified-charge combustion process developed in 1893 by the German engineer, Rudolf Diesel. In the Otto engine, heat input (spark ignition) is provided at a constant volume, while in a Diesel engine, compression ignition occurs under constant pressure. Both the Otto and Diesel cycle engines are internal combustion engines that ignite the fuel inside the cylinder. The more efficient Stirling cycle engines utilize an *external* combustion process, whereby the source of heat or combustion takes place outside the cylinder of the engine. The external heat is transferred through a heat exchange material to a working gas (such as hydrogen, helium or liquid sodium) that is sealed inside the cylinder of the Stirling engine. As the working gas is heated, it expands and, in so doing, it causes the piston to move. This is why a Stirling engine is omnivorous, in that it can use virtually any form of energy or fuel that creates heat, including solar energy.

The early Stirling engines were considered to be very reliable, but they were also relatively heavy, which caused them to have a low weight-to-power ratio. Internal combustion engines have a power density of about 4 to 5 pounds per horsepower in contrast to about 13 pounds for a Stirling engine. That is why they lost the dollars-per-horsepower race to the Otto and Diesel engines. In addition, the early Stirling engines were relatively complicated because the external combustion required the heat to be transferred through heat exchangers and ducts which took up half of the volume of the engine. As a result, while major automobile manufacturers like General Motors and Ford continued to research Stirling engine designs, they elected to utilize Otto and Diesel engines for their automotive vehicles.

While Otto and Diesel cycle engines are well-suited for transportation vehicles, they are not very appropriate for solar dish gensets because of their internal combustion characteristics. What is required is an engine that operates on external heat, such as a Stirling, Rankine or a Brayton (gas turbine) cycle design. After the Arab oil embargo of 1973, there was a serious attempt in the U.S. to investigate the most cost-effective solar technologies which could be developed. As a result, several dish test-bed systems were constructed for NASA and the Department of Energy with all three of the engine designs[13]. It became clear from the field tests that the Stirling engine power conversion units were the most promising in terms of efficiency, long life and cost.

There are two basic types of Stirling engines: Kinematic and Free-piston. Kinematic configurations were initially used in the solar genset tests, and they are similar to conventional internal combustion engines in that they have a crankshaft that is driven by pistons. For that reason, most automobile manufacturers were developing Kinematic-type Stirling engines. Free-piston Stirling engines, on the other hand, do not have a crankshaft, valves or any physically connected moving parts that are prone to wear and corrosion. They have only two moving parts whose motion is determined by their respective masses, as well as gas bearings (i.e., there is no physical connection between moving members). The primary components include a displacer and power piston that are supported by hydrostatic gas bearings with noncontacting clearance seals[14]. The integration of the free-piston engine with a linear alternator (for electricity production) is schematically shown in Figure 5.20.

When heat is applied to the free-piston Stirling engine, the working gas expands which causes the piston to move back and forth. Because of the simplicity of the free-piston engines, they have the potential for an extremely long life (30 years), high reliability, and importantly from a solar-genset perspective, maintenance-free operation. Free-piston Stirling engines are quiet, hermetically-sealed and literally welded shut, making them air-tight. This means it would be impossible for dust or other particles to enter the interior of the free-piston Stirling engine. Most importantly, because of their simplicity, the free-piston Stirling engines are expected to be the least expensive to manufacture. For all of these reasons, the optimum solar technology at present is a dish concentrator integrated with a free-piston Stirling engine.

Although significant progress was made in the development of the solar dish systems, the Reagan Administration decided that such developmental research should be primarily left to the private sector. As a result, the initial prototype dish gensets pictured in Figure 5.13 and the entire JPL test-site at Edwards Air Force Base were dismantled. The Department of Energy transferred responsibility for research on solar dish systems to Sandia Laboratories in New Mexico, but funding for the already small solar thermal program was nearly discontinued. When Federal research funds were cut, McDonnell Douglas discontinued its solar thermal division that developed the dish systems pictured in Figure 5.14 and sold the rights and hardware to Southern California Edison.

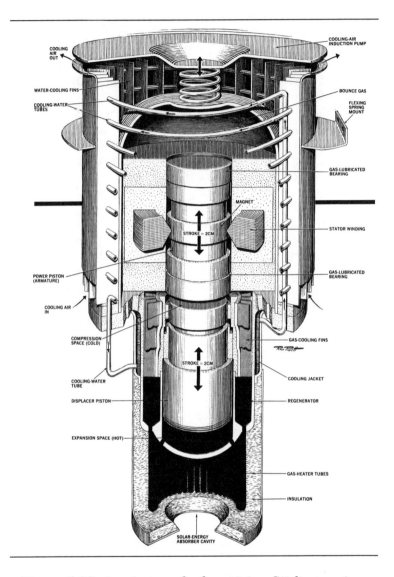

Figure 5.20: A cutaway of a free-piston Stirling engine-electrical generator system developed by Dr. William T. Beale, College of Engineering, Ohio University[15]. (Reprinted from *Popular Science* with permission © 1985 *Times Mirror Magazine)*

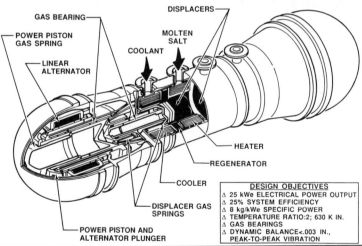

Figure 5.21: A dual cylinder free-piston Stirling Engine. The engine pictured above was developed for NASA by Mechanical Technologies, Inc. (MTI). The engine has a 25 percent overall efficiency and a 25 kilowatt power output. (Photo and cutaway courtesy of *MTI*, Latham, New York)

Southern California Edison, in cooperation with the Electric Power Research Institute (EPRI), continued to test the McDonnell Douglas solar dish system equipment until 1988 when it also discontinued its solar thermal research program. However, Southern California Edison did not terminate the solar dish program because the systems failed to operate successfully. On the contrary, research engineers from the utility concluded that both from a technical and economic perspective, the dish systems were one of the most attractive energy systems currently available. The real problem is that Southern California Edison is not planning to build any new electrical generating facilities for the foreseeable future. This is primarily because increasing numbers of large industries that were large users of utility-generated electricity are now generating their own power on-site with natural gas-fueled cogeneration systems.

What is significant is that utility-oriented engineers and planners are generally only interested in electricity production and not normally involved in hydrogen energy engineering research. In fact, virtually none of the dish genset engineering teams were even aware that the *International Association for Hydrogen Energy* even existed. It is equally significant that the hydrogen engineering community has been essentially unaware of the solar Stirling dish genset technology and its potential for large-scale hydrogen production. This is a classic case of a significant information-gap between highly-trained specialists who must, by necessity, focus their efforts in relatively narrow fields. The problem is compounded by the fact that there are few comprehensive energy research specialists who take a look at the big picture with a long-term view.

Notable exceptions are Dr. Glendon Benson and his engineering research and development colleagues who prepared a paper, "An Advanced 15 kW Solar Powered Free-Piston Stirling Engine." Their paper was delivered in 1980 at the 15th Intersociety Energy Conversion Engineering Conference held in Seattle, Washington, and later published by the *American Institute of Aeronautics and Astronautics*[16]. It was Benson and his colleagues who first purposed using solar powered Stirling dish gensets for large-scale hydrogen production. Hydrogen would ideally be generated with a high-temperature electrolysis system that has been under development by investigators at Brookhaven National Laboratory and Westinghouse Corporation[17]. Such a system could be integrated into the solar receiver that contains the fireball of hot air, and when it was optimized, overall system efficiencies in excess of 60 percent were predicted. This was in contrast to the maximum efficiency of 28 percent for silicon-based photovoltaic cells, but in spite of the significance of the announcement, the paper by Benson and his colleagues went unnoticed by the media and the international energy engineering community.

*Dish Genset Cost Estimates*

The cost per solar genset can be estimated from a detailed heliostat production cost study, "Heliostat Production Evaluation and Cost Analysis," that was prepared for the Department of Energy by General Motors Corporation (Warren, Michigan) in 1979[18]. Large numbers of heliostats (refer to Figure 5.22) would be used to concentrate solar energy on a large central receiver, referred to as a "power tower," (pictured in Figure 5.23). Although a power tower is like a giant dish concentrator, it has two significant disadvantages that smaller dish genset systems do not have: cosine losses and greater site specificity.

Power tower systems suffer from cosine losses, which simply means that the power output peaks at noon, and is substantially less in the early morning and toward the end of the day when the Sun is at a steep angle relative to the large array of heliostats. Smaller dish gensets do not have this problem because their small power conversion units are always held directly perpendicular to the Sun. Power towers are also very site-specific, which means they can usually only be constructed in relatively flat areas so the large array of heliostats can be accurately focused on the receiver tower. The smaller dish gensets, on the other hand, can be installed and efficiently operated on hilly or uneven surfaces. For these reasons, dish gensets will probably be more more cost-effective than power tower systems. However, the heliostats are, in effect, a flat mirrored dish in terms of engineering mass and overall production costs. As such, their cost estimates can be directly extrapolated to dish concentrator systems.

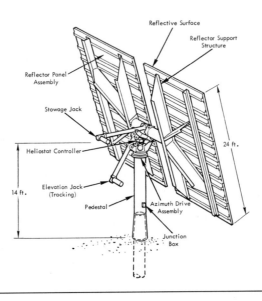

Figure 5.22: An individual heliostat designed by General Motors Corporation.

The G.M. study determined that if only 250,000 heliostats were produced annually, the factory cost per unit cost would be about $3,302. When profit and installation costs were included in the calculations, the per unit cost was increased to $4,390. Unlike a heliostat, a dish genset system must have an individual power conversion unit (i.e., an engine-generator set) in order to generate the electricity and/or hydrogen. According to Benson and his colleagues, an advanced free-piston Stirling engine with a 100,000 hour life would be expected to cost about $60 per installed kilowatt (kW), assuming 100,000 units were produced annually[19]. At this rate a 20 kW-output Stirling power conversion unit would be expected to cost about $2,000. This means the total cost of an installed 20 kW Stirling dish genset would be about $6,400 in 1979 dollars, which would be about $10,000 in 1990 dollars, assuming annual inflation rates of 4 percent.

Figure 5.23: Multiple heliostats form a large point-focus concentrator called a solar "power tower."
(Courtesy of Southern California Edison Company)

What is not yet known is how much the per unit cost would decrease if 5 or 10 million units were produced annually, particularly if the lower-cost LaJet bicycle-type concentrator system is used. It is worth noting a typical automobile engine that has a wholesale cost of around $1,000 and generates about 130 horsepower, has an electrical potential of about 96 kW. Thus, the installed cost per kW is only about $10, which means a 20 kW output engine would be expected to cost about $200, rather than the $2,000 cost that is assumed if only 100,000 Stirling engines are produced annually. This is only one-tenth the cost of a unit produced in relatively small quantities, and if similar cost reductions are possible for the entire system, the $10,000 cost could be reduced to under $2,000 for a 20 kW unit. That would mean the installed capital cost would be about $100 per installed kW, but because the

Sun only shines about 8 hours a day, it is necessary to multiply the $100/kW times 3, resulting in capital costs of about $300 per installed kW. This is only a fraction of the capital costs that are required for large centralized fossil fuel or nuclear facilities, which range from $1,500 to $2,500 per installed kW. Moreover, *solar gensets will have no fuel costs, and they will not generate any radioactive or other toxic by-products in their operation.* As a result of all of these factors, when solar genset systems are optimized, they should be able to generate electricity for roughly $0.01 to $0.02 per kWh, which means liquid hydrogen could be generated at a cost that is equivalent to gasoline costing $1.65 to $2.10 per gallon.

## Line-Focus Systems

At present, roughly 90 percent of the electricity generated by solar energy does not come from dish gensets or photovoltaics, but line-focus systems developed by Luz International Limited, a solar engineering firm headquartered in Los Angeles, California. Unlike point-focus concentrator systems that focus the Sun's energy into a small circular aperture area, line-focus reflector assemblies focus the solar radiation onto specially coated steel pipes which are mounted inside vacuum-insulated glass tubes. Inside the heat pipe, that runs the length of the trough-like solar concentrator, is a noncorrosive heat-transfer fluid, such as oil, that is heated to about 735 degrees Fahrenheit. The hot oil is then circulated through a heat exchanger, which then heats water until it becomes superheated steam that can then be used to power a conventional steam turbine. Luz has signed a long-term contract to provide electricity to Southern California Edison Company, and as a result, natural gas is used about 30 percent of the time as a back-up to the solar component of the Luz energy system.

What makes the Luz line-focus system significant is not that it works, but that even on a relatively small scale of production, it has been able to generate electricity at substantially less cost than new nuclear power plants. Line-focus solar technologies have been known about for years, but the cost of electricity from such systems was projected to be around $0.25 per kWh, which is roughly 4 times more expensive than the electricity generated from fossil-fueled utility systems whose electricity costs average about $0.06 per kWh. But the engineers at Luz, working in cooperation with solar research engineers at Sandia National Laboratory in New Mexico, were able to progressively increase efficiencies while reducing capital costs. With such improvements, the resulting costs of electricity have been reduced to about $0.08 per kWh, and Luz engineers are confident that they will eventually be able to generate electricity at $0.06 per kWh, and perhaps less.

Figure 5.24: A state-of-the-art line-focus concentrator system located in Kramer Junction, California. (Courtesy of Luz International Limited)

At $0.06 per kWh, the resulting cost of liquid hydrogen would be about $3.80 per gasoline equivalent. But if line-focus systems were going to be built on a scale sufficient to produce hydrogen as a substitute for fossil fuels, their scale of manufacturing would have to be dramatically increased. That, in turn, would further reduce the cost of the electricity or hydrogen that is generated.

*Solar Economics*

The miracle of mass production revolves around the fact that in most cases, if more individual units of something are produced, the cost of each unit will decrease accordingly. It is important to realize that when most energy analysts calculate the relative energy costs that would be generated from solar technologies, they make assumptions about how many units will be produced per year. In most cases, there is a general assumption made that solar technologies would at best provide only a small percentage of electricity production, and electricity only accounts for about 15 to 20 percent of the overall energy use of most industrialized countries. As a result, the projected production rates for solar genset systems were always assumed to be only a few thousand units per year.

If, however, solar technologies and related systems are to provide sufficient electricity to produce enough hydrogen to displace the use of fossil fuels, in excess of 8 to 10 million genset units per year would need to be produced over a period of many decades -- just for the U.S. market! This profoundly affects the economics because it means totally committed factories could be designed and built to mass-produce the renewable energy equipment. This, in turn, means the per unit cost will decrease substantially. Solar technologies are also modular in design (i.e., small individual units can be connected and thereby integrated into a larger network or system). This modular nature of renewable resource technologies is important for the following reasons:

1. As new technology is developed to further reduce the cost of producing hydrogen, it can easily be assimilated into a modularized energy system which inherently lends itself to an era of rapid technological change. This is in contrast to nuclear power plants, which can take over 10 years to build. This means that from the time nuclear plants are initially designed to the time they are finally completed, it is likely that they will become economically and technologically obsolete.

2. Energy systems that are modularized lend themselves to mass-production; they can be purchased incrementally and will have construction lead-times of days or hours. This is in contrast to nuclear plants that require billions of dollars to be financed for the 10 or more years it takes to build the facility. This is an important financial consideration, given the direct relationship of time and money.

3. Modularized energy systems are inherently more reliable than single-source centralized facilities. If one elephant could do the same job as one million ants, which would be better to use? Consider that if the elephant dies or gets sick, the work cannot be accomplished. But if one has a million ants working and several die or are otherwise unable to work, the overall job still gets done. Simply put: *There is safety in numbers.*

Speaking of numbers, it is of interest to calculate how many solar gensets would have to be built and installed in order to make the U.S. energy independent from the remaining fossil fuel reserves. In 1986, for example, the U.S. used an estimated 64.2 quads (i.e., quadrillion) Btu of energy. Recall that a quadrillion is a 1 followed by 15 zeros (1,000,000,000,000,000). One Btu is roughly equal to the amount of energy contained in one match. One gallon of gasoline contains about 119,000 Btu. Electricity is measured in watts, and one watt of electricity operating for one hour is equal to about 3.4 Btu. A thousand watts of electricity is a kilowatt (kW), and when one kilowatt operates for one hour (kWh) it is equivalent to 3,413 Btu.

If a solar genset has an average net peak output of 20 kW, it would be generating 68,260 Btu per hour of direct sunlight. Assuming an average of 8 hours of sunlight per day, multiplied by an average of 300 days per year, a total of 163.8 million Btu per year would be generated by one 20 kW solar genset located in the American Southwest. Given this information, simple division will give the approximate number of how many 20 kW solar gensets would have to be built in order to equal current U.S. energy production (i.e., about 65 quads):

$$\frac{65 \text{ Quadrillion Btu}}{163.8 \text{ Million Btu}} = 396,825,000 \text{ Solar Gensets}$$

Therefore, to provide enough energy (in the form of electricity and hydrogen) to displace all fossil fuel and nuclear production in the U.S., about 400 million solar gensets will need to be manufactured and installed. Note that the 20 kW output solar genset could be a point-focus concentrator dish system, a photovoltaic array, or a wind machine. Assuming the solar gensets would be expected to cost from $2,500 to $10,000 each (depending on how many units are produced annually), the overall cost of building 400 million units would range from 1 to 4 trillion dollars. It is important to note, however, that most of the investment will come from private sector investments rather than from deficit-financed tax dollars.

*Return On Investment (ROI)*

While nuclear weapons may provide some measure of defense, they do not provide an economic rate of return on the invested capital. Indeed, once built, additional money and manpower must continually be expended to maintain such complex and dangerous weapons systems. This is why only governments will buy such equipment. Solar gensets and other renewable resource technologies, by comparison, will provide an economic rate of return on their investment. In addition, their relative risk factors are small in comparison to radioactive nuclear systems that have no financial bottom line; or multi-billion dollar offshore drilling rigs that have been lost in storms; the increasing numbers of ever more expensive dry holes that the oil industry keeps drilling; or the ecological disasters that can result from oil spills and strip mining.

These are some of the main reasons why solar technologies will provide a relatively low-risk rate of return on their invested capital, and as a result, minimal government expense. While some people might automatically assume that the major oil companies would seek to prevent the use of solar-hydrogen energy systems, in fact, it has been the major oil companies that have helped to sponsor many of the technical conferences that are put together every two years by the *International Association for Hydrogen Energy.* This should not be surprising. Senior executives of the major oil companies have repeatedly indicated that they and their stockholders are always looking for legitimate long-term alternative energy investments, especially if the renewable energy systems can provide them with a low-risk, renewable rate of return on their invested capital.

Oil executives know better than anyone about the difficulty of finding new oil. This explains why they are now finding it is less expensive to simply buy their competitors' oil reserves rather than to find what is left. Since 1980, there have been more than 325 mergers and divestitures within the oil industry worth more than $62 billion[20]. Rather than oppose a transition to renewable solar-hydrogen production technologies, the oil industry will likely be at the heart of their mass-production when the appropriate national energy policy has been established. The oil companies will eventually evolve into hydrogen companies as they increasingly shift their investment capital to the relatively low-risk renewable resource technologies that will be manufactured by automotive and aerospace-type industries.

## U.S. Energy Policy

Traditionally, the energy policy of the U.S. government has revolved around encouraging private energy companies to explore and extract fossil fuels and other natural resources. As a result of the Arab Oil Embargo that occurred in 1973, the Nixon Administration established a goal of making the U.S. energy-independent from foreign energy suppliers without specifying how the objective was to be accomplished. The Carter Administration called for the "moral equivalent of war" to be declared in order to make the U.S. less dependent on energy imports.

But although the research into solar technologies was substantially increased during the Carter Administration, the bulk of the research dollars involved developing other fossil fuel resources, principally oil shale and coal.

With the election of Ronald Reagan, U.S. energy policy shifted back to finding and consuming what is left of the fossil fuels and building additional nuclear power plants. President Reagan abandoned the national goal of energy independence established by the Nixon Administration. Thus, what little engineering and financial resources that were being focused on the renewable resource problem were shifted to an already substantial arms race, which included a new range of highly complex and expensive "Star Wars" space weapon systems. The only reason the solar thermal program was continued is because it looked like the most cost-effective technology that could power a space station and the space-based weapons systems.

The members of Congress and the advisors to President Reagan apparently did not realize that their energy and environmental policies were an even greater long-term threat to the security of the U.S. than the Soviet armed forces. This is because an armed conflict with the Soviet Union is unlikely, whereas the problems of resource depletion and environmental contamination are very much in the process of happening. Given the exponentially worsening global environmental problems, and the long lead-times and substantial capital investments that are required for making major industrial changes, it is especially unfortunate that there has not been an effort to shift the limited U.S. financial and engineering resources from the arms race, which neither the U.S. or the Soviet Union have been able to win, to the renewable energy resource race that would directly benefit all countries.

In that regard, the Soviet Union needs to make an industrial transition to renewable resources as much as the U.S. does. The Soviets have been the largest oil producers in the world for many years, but their oil production is reported to have peaked in 1984 and is expected to continue declining. As a result, it is in the national interest of both superpowers to undertake a transition to renewable resources before the diminishing fossil reserves eventually trigger an armed conflict over the remaining Middle East oil reserves. From a U,S. perspective, the possibility of a war in the Middle East with the Soviet Union is an especially serious concern because the U.S. is not in a strong strategic position in that part of the world. In addition, most military analysts do not believe U.S. forces could successfully neutralize Soviet forces without the use of nuclear weapons. This, in turn, could rapidly escalate into the full-scale nuclear exchange that has been dreaded for so long by so many.

If the solar gensets end up costing $10,000 each, and a total of 400 million are manufactured to make the U.S. energy independent, the $4 trillion investment would be substantial, but manageable. Hopefully, the per unit cost can be reduced to $2,000 or $2,500, which would reduce the investment to about $1 trillion. But if this amount of money seems unreasonably large, it is well to remember that the renewable resource solar-hydrogen technology investments will likely be spread over a period of four decades. Thus, annual industrial investments of from 20 to 100 billion dollars would be required, depending on the final cost of the renewable resource systems.

To put these numbers into perspective, consider that more than $300 billion *deficit-financed* tax dollars are currently being spent annually on the U.S. Department of Defense alone. At that rate, $4 trillion will be spent in only 13.3 years. In sharp contrast, the vast majority of the renewable resource investments are going to come from the private sector, and not deficit-financed taxpayer dollars. This is because solar-hydrogen and other renewable resource technologies will provide a relatively low risk, *renewable rate of return.* As a result, government financing will be minimal. However, the Federal and State Governments do have an important role to play in terms of establishing a national energy policy that will be oriented around a transition to renewable resources.

Like the war effort in World War II, a cooperative effort between industry and government is going to be required if such a fundamental energy and industrial transition is to occur. If, for example, 10 million solar dish gensets were produced every year, which is roughly equivalent to current U.S. automobile production, it would take about 40 to 50 years to build and install all 400 million solar gensets. This timetable underscores the need to begin quickly and to act with a sense of urgency, given the fact that existing U.S. oil reserves will be exhausted in only 10 to 15 years, and current world oil reserves, most of which are in the Middle East, are expected to be seriously depleted in 35 to 40 years.

In addition, long before the known oil reserves are exhausted, the price per barrel will increase substantially. Whether or not the U.S. is going to have long-term economic stability will depend upon how quickly it can mobilize its national resources -- which are substantial -- to change course. It is important to shift from the arms race to the energy technology race before the cost of doing so increases further. If solar gensets, OTEC plants and wind energy conversion systems were developed in the 1950's and 1960's, instead of nuclear-fission facilities, the average cost to build such systems would have been substantially less than if the same facilities were built today. This is because the price of oil has gone up substantially since 1950, and as a direct result, so has the cost of everything else. All that has been accomplished by waiting is the price of everything has increased.

It is also interesting to note that if the industrial transition to renewable solar hydrogen resources and technologies had started in the 1950's, enough solar technologies could have been built to have made the U.S. energy independent by 1973, when the first Arab oil embargo took place. In addition, all of the energy pipelines, ships and other infrastructure technologies would already be hydrogen-compatible, and much of the environmental devastation involving oil spills, radioactive waste, the production of greenhouse gases and acid rain could have been avoided. Although many U.S. Government policy planners did take seriously the warnings of Dr. M. King Hubbert and the other geophysicists at the U.S. Geological Survey concerning the eventual depletion of U.S. oil reserves, rather than concentrating on the development of solar technologies, billions of taxpayer dollars were spent instead on developing nuclear-fission facilities.

One of the primary explanations as to why the high-risk nuclear option was favored over the simpler solar technologies was that nuclear advocates initially believed that nuclear systems would be able to produce electricity "too cheap to meter." In addition, there were many people in the U.S. government who wanted to provide a more positive public relations image for nuclear technologies and to justify the vast expenditures of money and engineering talent that went into the development of nuclear weapons. The result was an "Atoms for Peace" program that was put forth in an attempt to provide a constructive peacetime use for the nuclear-fission technologies. As it turned out, this was a tragic miscalculation, because unlike nuclear energy systems, the renewable solar-hydrogen resource technologies could have been optimized and mass-produced with the automotive and ship-building technology that existed at the time.

*Conclusions*

Solar technologies provide a realistic method of providing the energy necessary to extract hydrogen from water. Although photovoltaic cells are a promising solar technology option, at present, the electricity they generate is far too expensive to produce hydrogen at a price that would be competitive with petroleum-based fuels like gasoline. Other solar technology options, however, such as point-focus concentrator "dish" systems, ocean thermal energy conversion (OTEC), or wind energy conversion systems could be mass produced to provide hydrogen fuel at a price that is competitive with fossil-based fuels.

The renewable resource energy industry will rival the automobile industry when production is in full-swing. The net result of such an industrial transition will be that essentially pollution-free industries will be producing pollution-free renewable-energy machines that could be utilized by the most advanced industrial nations or the most remote villages in Third World countries. From the consumer's perspective, there is no reason to be concerned about which renewable resource technology will ultimately turn out to be the most cost-effective. May the best engineering teams win. As more improvements are made, it is reasonable to assume that the energy costs will be continually reduced, in contrast to the non-renewable and nuclear fuels, which will become increasingly expensive as their resources dwindle.

With respect to the the arms race, there is the hard reality that the defense industries provide a substantial economic underpinning for both the U.S. and the Soviet Union. This being the case, if the arms race were to actually be resolved, it would initially have a serious economic impact on both countries because millions of people would lose their defense-related jobs. As a result, the arms race has a substantial economic force behind it. President Eisenhower understood this problem, which he referred to as the "military industrial complex," and he warned that it was developing into an institution that was growing out of control. The significance of an industrial transition to renewable resources is that it represents a formula that could allow many of the defense-related industries to make an orderly transition to manufacturing renewable resource technologies instead of weapons.

Fundamental problems require fundamental solutions. If the U.S. establishes a national policy to make an industrial transition to renewable resources, other countries will follow suit because it is in their own self-interest to do so. It is important that the transition to renewable resources occur as soon as possible because everything will only get more expensive over time, and because critical decisions are being made to lease what is left of the remaining wilderness areas to oil and other energy companies so they can extract the last of the fossil fuels and uranium. The result of these shortsighted decisions will be to leave nothing for those who come after us, except for ruined land, contaminated water and toxic chemicals, including the radioactive wastes that will seep and spread for centuries. Perhaps the greatest tragedy of all is that the incalculable environmental damage that has already occurred did not need to happen. Nor is it necessary to stand by and let it continue.

The renewable resource technologies reviewed in this chapter document the fact that there are many viable alternatives to the existing fossil fuel and nuclear energy systems. Chapter 6 will review the physical resources that will be required if such technologies are to be implemented on a global scale.

Chapter 6

# RENEWABLE ENERGY RESOURCES

*An Overview*

In the previous chapter, some of the most technically and economically viable solar technologies that could be used for large-scale hydrogen production were discussed. In this chapter, the energy potential of these options will be outlined, as well as the land and water resources that would be required for their large-scale implementation. To put the energy potential of renewable resources into perspective, the following information is useful:

1. According to the U.S. Energy Information Administration (Washington, D.C.), world energy production in 1986 was about 314 quadrillion (quads) Btu. Roughly 277 quads (or about 88 percent of the world total) was generated by fossil fuels, 21 quads (or about 7 percent) was generated by hydroelectric dams, and about 16 quads (or about 5 percent) was generated by nuclear power plants[1].

2. Total U.S. energy production in 1986 was about 64 quads, which is about 20 percent of the world total. Over 56.6 quads (or about 88 percent) of the U.S. production came from fossil fuels. Nuclear provided 4.4 quads (or about 7 percent), and hydroelectric dams provided 3 quads (or about 4.6 percent)[2].

3. A megawatt is equal to 1000 kilowatts, and a kilowatt is equal to 1000 watts. A kilowatt-hour is equal to 3,413 Btu. Thus, a megawatt-hour is equal to 3,413,000 Btu.

The amount of solar energy that is received by the Earth each year has been estimated to be in excess of 5 million quads. Although roughly one-third of this solar input is reflected back into space by the Earth's atmosphere, that still leaves over 3.5 million quads of input energy annually. This is more than 11,000 times the 314 quads of energy that were consumed in 1986 by humans worldwide. From an energy production perspective, sunlight carries energy that is equivalent to about 1000 watts per square meter of exposed surface. An important question is: *How is this vast resource distributed with respect to solar energy conversion technologies?*

### Wind Energy Resources

Wind is defined as the movement of air across the Earth's surface. Winds are primarily the result of the Earth's atmosphere absorbing the immense quantities of energy produced by the Sun. The differential heating of the Earth's surface causes a lateral heat flow that keeps the gases and particles of the global atmosphere in motion. Other winds are caused by what is referred to as the "evapotranspiration cycle," which involves heat being stored in the atmosphere until it rains, cooling things off, thereby allowing the heat build up to start over again.

The principal supply of "fuel" for the Earth's heat engine is in the form of water vapor (which happens to be the primary combustion product of hydrogen and oxygen). Water vapor is a kind of fuel storage system because as it is alternately evaporated and condensed, energy is released to the atmosphere, and the most evident form of this energy is the wind. Although wind energy varies from second to second and is not equally distributed around the globe, for centuries sailors have known that it is remarkably consistent over time within large areas. In general, the major wind systems intensify from equator to pole, with much modification of these patterns due to the relation of land and water masses and topographic features such as mountains, trees or buildings.

The World Meteorological Organization at Geneva, Switzerland, surveyed the world's existing data of wind resources and concluded that there was an electrical potential of about 20 million megawatts[3]. Given that there are 8,760 hours in a year, and assuming the average wind machine could achieve a 40 percent operating capacity, over 70 billion megawatt-hours of electricity could be generated annually. If the wind machines were to have a 60 percent operating capacity, they would be able to generate in excess of 105 billion megawatt hours annually. This is in contrast to the 92 billion megawatt-hours (i.e., 314 quads) of energy consumed by the total global human community in 1986.

While it is true that placing large numbers of wind machines on mountain tops is not an environmentally acceptable alternative, there are other options discussed in the previous chapter that include placing large numbers of wind machines out at sea or utilizing advanced vortex generator configurations that do not need to be placed in mountain top environments. There is no question about the vast potential of wind-powered energy conversion systems. One can only wonder why so many of the industrialized nations have neglected this renewable resource technology for so long.

### OTEC Resources

The oceans contain 98 percent of the Earth's water, and they make up over 70 percent of the Earth's surface area that receives solar radiation. This makes the oceans the largest solar collector on the Earth, and it has cost nothing to build. Moreover, half of the Earth's surface lies between the latitudes 20 degrees North and 20 degrees South, which is mostly occupied by the tropical oceans where ocean thermal energy conversion (OTEC) plants could efficiently operate.

According to calculations by Dr. Clarence Zener, Professor of Physics at Carnegie-Mellon University, the potential energy that could be extracted by OTEC plants located in the tropical ocean areas would be about 60 million megawatts[4]. Dr. William Avery, director of the Applied Physics Laboratory at Johns Hopkins University, has calculated the OTEC potential to be about 10 million megawatts[5]. Assuming the lower value is correct, and assuming the OTEC systems would have an operating capacity of about 80 percent, they would be able to generate about 70 billion megawatt-hours per year. If Zener's more optimistic calculations are correct, the OTEC systems could generate over 400 billion megawatt-hours per year. (Recall that the total human energy consumption in 1986 was 92 billion megawatt-hours.)

Thus, both wind and OTEC systems could, in and of themselves, generate enough electricity and/or hydrogen literally to run the world -- without using any of the Earth's remaining fossil fuel reserves. It follows that all of the impending environmental disasters that will result if those remaining fossil fuels are extracted, shipped and burned, could be avoided. In addition, Zener and his colleagues have calculated that even if 100 percent of the world's energy needs were provided by OTEC systems by the year 2000, and even assuming the entire world was consuming energy at the rate that the U.S. does, the surface temperature of the tropical oceans would only be lowered by less than 1 degree Centigrade[6]. For all of these reasons, wind and OTEC energy conversions deserve careful consideration in formulating a national and international energy transition to renewable resources.

*Solar Genset Resources*

Although offshore wind and OTEC systems do not have requirements for large real estate or water acquisitions, that is not true for the predominantly land-based photovoltaic or Stirling powered solar genset systems. While solar energy may fall on most of the globe, it is not uniform in its distribution. Just as fossil fuel resources have been concentrated in certain geographical areas, the same is true for solar energy resources. A quick analysis of Figures 6.1 and 6.2 will confirm this fact. While this will be an unpleasant thought for those who are hoping that the large-scale use of solar technologies will put an end to the era of large international energy companies, the stockholders of the energy companies will be pleased.

Figure 6.1 provides an overview of the high solar insolation areas of the U.S. Note that solar radiation is measured in kilolangleys per square centimeter per year on a horizontal surface at ground level. A kilolangley is 1,000 langleys, and a langley is one calorie of radiation energy per square centimeter. This is equivalent to .216 kilowatt-hours (kWh) per square foot or 603 million kWh per square mile for every 200 kilolangleys. Note that the desert areas of the American Southwest, which cover vast areas of Arizona, California, Nevada, and New Mexico, receive the highest concentrations of solar energy in North America. Solar-hydrogen energy systems located in these desert areas will offer substantially higher returns on invested capital than if the same equipment were situated in a more northern latitude.

Most of the 200 million acres that make up these desert regions are uninhabited. Although numerous Indian reservations are located in these desert areas, it is highly likely that many of the Native Americans would welcome the renewable income that would be generated from installing large forests of solar-hydrogen systems. Assuming twenty 20-kilowatt peak output solar gensets are installed per acre, about twenty million acres (which is about 10 percent of the total area) would be required to install the roughly 400 million gensets and the related subsystems that would be necessary to produce the current annual U.S. requirements of 64 quads of energy.

Care would obviously need to be exercised as to where the vast forests of solar gensets would be placed. Much of the Sonoran Desert, for example, has a relatively fragile ecosystem in contrast to the Mohave Desert in California or the salt flats in Utah. From a solar input standpoint, it would be both desirable and theoretically possible to place all 400 million gensets within the southern portion of Arizona. However, given the many economic, environmental, political, and strategic-military reasons, it is reasonable to assume is that after a careful study of potential land areas has been undertaken, the solar gensets would be distributed throughout various optimal regions of the South and Southwest.

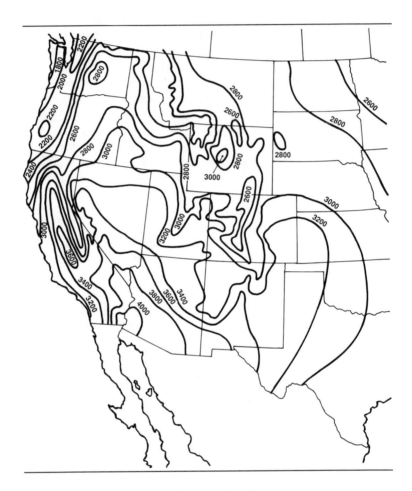

Figure 6.1: U.S. Solar Resources.
Annual Mean Total Hours of Sunshine in Langleys.
Source: *Atlas of The United States*, U.S. Department of
Commerce, June 1968. (Courtesy of Climatology
Laboratory, Arizona State University)

Once major automotive manufacturers such as General Motors, Ford and Chrysler are mass-producing solar gensets, individual home owners and companies from all over the country will want to purchase their own systems and become energy-independent. But for millions of urban dwellers who live in large high rise cities such as New

York or Chicago, they have neither the space nor the appropriate weather conditions for solar systems. As a result, most individuals in these areas will continue to purchase energy from centralized suppliers.

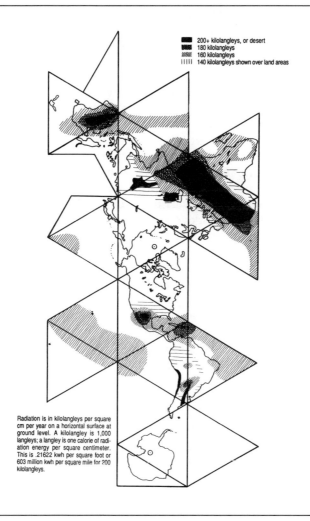

Figure 6.2: Global Solar Resources.
Mean Annual Solar Radiation. Source: *Physical
Geography: Earth Systems,* 1974.

The fact that solar land resources are concentrated in relatively limited areas of the U.S., and, in a larger context, in only limited areas of the planet, ensures that large-scale solar use will allow existing energy companies to maintain their strong economic position. This is because if a large energy company is going to be installing solar gensets by the millions for base-load solar-hydrogen production, it will obviously choose to place the gensets in areas with the highest levels of solar energy to maximize its capital investment. As a result, states like Arizona and California are like Saudi Arabia or Kuwait in terms of their solar energy resource. Unlike oil, however, the Sun's energy will be able to provide energy companies with a renewable rate of return on their investments.

The potential economic implications of large-scale solar development to states like Arizona, California, Nevada, New Mexico and Texas, are profound. It is interesting to note that the Federal government owns most of the mountains, while the state governments control most of the flat land areas where the solar gensets will be placed. As a result, state governments will be able to negotiate long-term lease agreements with energy companies in the same way that grazing rights for animals are negotiated. It is also significant that solar gensets only use a relatively small surface area of the land, which means the land could be co-used for grazing or agriculture. Countries other than the U.S. with substantial land and solar resources include Mexico, Saudi Arabia, Egypt, Israel, China, Australia and the Soviet Union.

Heavily industrialized countries which have neither the land availability or high-intensity solar input include Japan, Germany, France, Great Britain, Sweden and Denmark. Hydrogen produced in one country can easily be shipped to other countries in much the same way that petroleum and natural gas are presently transported by pipeline, trains, trucks and ships. Although the cryogenic tanker manufactured by General Dynamics in Figure 6.4 was designed to transport liquid natural gas, essentially the same type of cryogenic vessels would be used to transport liquid hydrogen from production facilities to world markets with great efficiency, and without environmental damage.

Figure 6.3:  A coastal liquid hydrogen fuel storage facility. Although this design was initially planned for storing liquid natural gas (LNG), similar cryogenic systems will be used for liquid hydrogen.

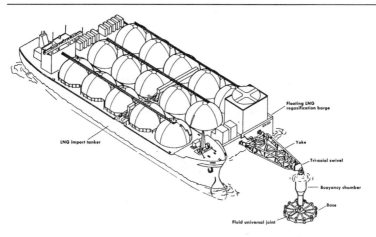

The floating LNG terminal was developed to be an entirely self-contained and self-sufficient operational facility.

Figure 6.4: A Cryogenic Tanker.  With liquid hydrogen transport ships, oil spills from accidents and regular tanker operations will no longer pose environmental problems.  (Courtesy of General Dynamics, Inc.)

As the industrialized countries make a transition to a hydrogen energy system, oil tankers will evolve into cryogenic liquid hydrogen transport ships that will not pollute the oceans during their regular operations. It is not commonly known that most of the oil that is released into the oceans is not due to accidents, but from normal operations where sea water is routinely used to clean excess oil from the tankers. This will not be necessary with liquid hydrogen transport ships. Although the fuel might be lost in the event of an accident, the liquid hydrogen would rapidly vaporize and dissipate harmlessly into the atmosphere. Thus, the ecological damage caused by routine tanker operations and accidents would be a thing of the past.

### Implementation Lead-Times

There are several key points to consider regarding implementation lead-times that are obvious when one examines Figure 6.5.

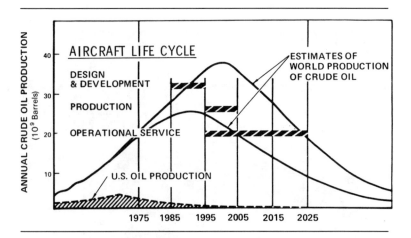

Figure 6.5: Life-Cycle Considerations.
Given current forecasts of oil availability, billions of dollars worth of aircraft are being designed to operate on a fuel that will not be available for the life of the aircraft.
(Courtesy Lockheed Aircraft Corporation[7])

Note that there are two estimates of world crude oil production in Figure 6.5. These two curves represent the conservative and liberal estimates of projected world oil reserves. It is also important to note the relative U.S. oil production, which peaked in 1970, is expected to be substantially exhausted around the year 2000. The graph in Figure 6.5 was prepared by senior Lockheed Aircraft engineers. The purpose of the illustration was to explain to congressional and other governmental planners that American industry is spending billions of dollars building airplanes that are designed to burn a fossil-based fuel (aviation kerosene) that will not be economically available for the 20 to 30-year life of an average commercial aircraft. As a result of this obvious problem, the Lockheed engineers proposed to members of the U.S. Congress and the Department of Energy that the U.S. should begin designing and building alternative-fueled aircraft that would not be dependent upon fossil fuels. Unfortunately, the U.S. officials did not take any action.

While Figure 6.5 is concerned only with aircraft, it is important to consider that pipelines, ships, trains, trucks and other long-term capital intensive industrial infrastructure investments are equally affected by the economic availability of fossil fuels. At present, most of the existing million-mile natural gas pipeline system in the U.S. could accept up to 20 to 25 percent gaseous hydrogen with little or no modifications, but to transport pure hydrogen will involve modifying existing pipelines or laying new ones. This underscores how important it is to have long-range strategic planning because billions of dollars can be saved if the pipelines that are going to be laid in the next 20 years are hydrogen-compatible. If such advanced planning is undertaken, when pure hydrogen is eventually phased in on a large scale in the future, the enormous investment in the energy pipelines and other infrastructure items will not have been wasted.

It should be clear that it is much more cost-effective if a long-term strategic energy plan and timetable is developed and understood by representatives of both industry and government. The failure to plan ahead will only result in crisis and panic reactions, conditions hardly conducive to rational thinking and planning.

*Water Considerations*

One gallon of water has a hydrogen energy content of about 52,400 Btu, compared with a gallon of gasoline, which has about 119,000 Btu. This means about 2.3 gallons of water will be required to extract enough hydrogen to equal the energy contained in one gallon of gasoline. In order to place the water issue in perspective, it is important to realize that it takes about 18 gallons of water to make a gallon of gasoline from crude oil. This is because oil refineries use large amounts of water to generate high-temperature steam, which is reacted with the crude oil to break or "crack" its long-chained hydrocarbon (hydrogen-carbon) molecules into the lighter molecules that make up isooctane fuels such as gasoline and aviation kerosene.

Water is a molecule made up of two atoms of hydrogen ($H_2$) that are electrochemically bonded to one atom of oxygen (O). The chemical symbol is $H_2O$. It is a simple matter to separate the hydrogen and oxygen in water with electricity, a process referred to as electrolysis. It is also possible to split water thermochemically, that is, with high temperatures (usually in excess of 1,600 degrees Fahrenheit). Such high temperatures can be achieved with both nuclear and solar sources, as the Gulf Oil advertisement in Figure 6.6 indicates.

Numerous major oil companies (most of whom like to view themselves as not just oil companies, but *energy* companies), have been involved in helping to find cost-effective ways of splitting the water molecule for hydrogen production. Oil companies have also actively supported hydrogen technical conferences organized by the *International Association for Hydrogen Energy*. This is primarily because hydrogen can also be made from nuclear sources, and many oil companies have made substantial investments in nuclear fuels. But in any case, while many people automatically assume that oil companies will try to oppose the transition to a renewable resource "hydrogen economy," the reality is quite different. Companies like Exxon, Phillips Petroleum, and Atlantic Richfield have, to a certain extent, already been involved in researching and developing various solar and hydrogen production systems, and when the governments of the world mandate that only renewable resource technologies will be acceptable, those will be the systems that the energy companies develop.

Figure 6.6: Hydrogen From Water. Note that the advertisement for the use of hydrogen energy was paid for by a major oil company. It appeared in numerous popular publications in 1984, including *Time* and *Newsweek*.

Although thermochemical methods of splitting water are promising, a far simpler method that has been used for years involves the process of electrolysis whereby an electrical current is passed through the water in a device called an electrolyzer. The electrolysis of water involves placing two electrodes, one positive and one negative, into a solution of water that has been made conductive by the addition of an electrolyte such as potassium hydroxide or sulfuric acid. As direct current (DC) electricity is applied, the water molecules will separate, with the hydrogen gas molecules being attracted to the negatively charged cathode while the oxygen gas rises out of the solution at the positively charged anode. Water is continuously added to replace the water that has been broken down into hydrogen and oxygen.

From a large-scale hydrogen production perspective, to generate 64 quads of hydrogen, about 3.8 million acre feet (maf) of water would be required each year. An acre foot contains about 325,900 gallons, and although 3.8 maf is less than 1 percent of the annual flow of the Columbia River, which separates Oregon from Washington, it is almost what the entire state of Arizona consumes annually. While desert areas tend to have plenty of the land and sunshine that are required for solar technologies, the one thing they do not have is water. Thus, any proposal for a water-intensive hydrogen production complex must have a realistic strategy for securing the necessary fresh water from sources other than existing rivers, aquifers, reservoirs or precipitation.

### Water Options

While conventional fossil fuel and nuclear facilities require large amounts of water to produce electricity (about one to two gallons per kilowatt hour of electricity generated), solar powered energy production systems will be able to produce electricity without *any* water requirements. As a result, the most obvious solution would be to make electricity in the deserts with large numbers of solar gensets, and send it to where the feed-water for hydrogen production is more abundant.

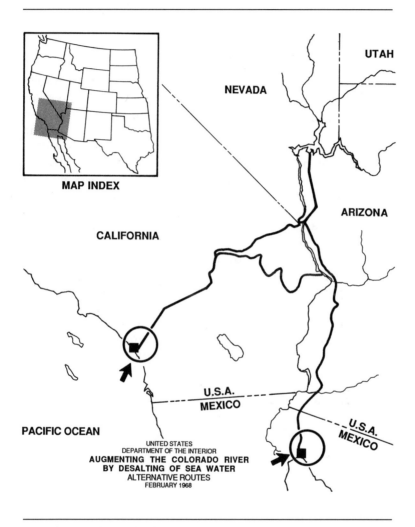

Figure 6.7: Pumping desalinated sea water into the desert. To provide the fresh water that will be required for large scale-hydrogen production within the arid desert regions, one option would involve building coastal sea water desalination facilities. The fresh water could then be transported with solar-powered pumping stations to the desert areas where it is needed.

Assuming the electricity that is generated by the solar gensets in the American Southwest is sent to sea water desalination facilities located in southern California or Mexico, the large-scale production of hydrogen could then take place in centralized electrolysis facilities. Although this option is technically feasible, it may actually be far more efficient to pump desalinated sea water to the solar gensets in the deserts. This is because solar dish gensets are able to incorporate an advanced, but highly-efficient high-temperature electrolysis process that has been under development by investigators at the Brookhaven National Laboratory and Westinghouse Corporation. This process is able to utilize part of the heat generated from the focal point of the solar dish to help break the water down into hydrogen and oxygen.

The use of high-temperature electrolysis can reduce the electrical energy required to separate the hydrogen from oxygen in water by one-third to one-half. This is why there are substantial engineering and economic incentives to take advantage of the 1,200 to 1,500 degree Fahrenheit "fireball" of hot air which is generated at the focal point of a solar dish system. While electricity would be needed to pump the water from coastal areas outlined in Figure 6.7, it is presumed that the pumping stations would also be using electricity generated by solar-powered genset systems. Preliminary calculations indicate that sending the water to the solar gensets will ultimately result in the least expensive method of generating hydrogen.


*NAWAPA*

In addition to sea water desalination, a more fundamental resolution to the entire Western water problem was proposed in 1964 by the Ralph M. Parsons engineering firm, located in Pasadena, California. Referred to as the North American Water and Power Alliance (NAWAPA), this project would involve diverting substantial amounts of fresh water (160 maf per year) from sources in Alaska and Canada where it is abundant (annual runoff exceeds 800 maf), to areas in Canada, the lower 48 states of the U.S. and Mexico, where it is not. (Refer to Figures 6.8 and 6.9.)

Roughly half of the water would go to the U.S. under the NAWAPA proposal, and the remaining 80 maf would be equally divided between the countries of Canada and Mexico. The NAWAPA project would easily provide the fresh water necessary for the proposed modularized multi-quad solar-hydrogen production facilities in the Southwest, thereby eliminating the task of having to pump it up from the Pacific Ocean. An even more compelling reason to incorporate the NAWAPA as a fundamental component of an industrial transition to renewable resources exists. It is the only proposal advanced that can enable the Western river systems of the U.S. to maintain the critical salt balance necessary for the long-term survival of Western agriculture. This serious salt accumulation problem has received little public attention, but if it continues for another 20 or 30 years, it will leave much of the richest agricultural region in the U.S. salt-encrusted and barren.

In a paper published in *Scientific American,* Dr. Arthur F. Pillsbury outlined the salt problem in great historical detail[8]. Pillsbury is especially well qualified to provide such insights. He served for many years as Chairman of the Department of Irrigation at the University of California at Los Angeles (UCLA). At the time of his retirement in 1972, he was professor of engineering and director of the UCLA Water Resource Center. In his paper, Pillsbury points out that many ancient civilizations rose by diverting rivers with a high salt content for the irrigation of their crops. Each one ultimately collapsed because as the water in the soil evaporated, it left behind the salt, which kept accumulating until the plants could no longer survive.

The most fruitful of the ancient systems was the Fertile Crescent region, a broad valley formed by the Tigris and Euphrates rivers in what is now Iraq. Six thousand years ago, the area was occupied by the Sumerians, who successfully diverted the river water to irrigate their crops. From there civilization spread eastward through what is now Iran, Afghanistan, Pakistan, India, and eventually China. At its peak of productivity each irrigated region is believed to have supported well over a million people. But all of those civilizations collapsed, and for the same reason: salt accumulation. Although floods, plagues and wars took their toll, in the end, civilizations based on irrigation faded away due to the accumulation of salt in their soil. Six thousand years later, the land in Iraq is still salt-encrusted and barren. If irrigated agriculture continues as it is presently, some of the richest farm land in the U.S. will also be lost to salt accumulation.

Figure 6.8: North American Water and Power Alliance.
(Courtesy of Parsons Engineering Company)

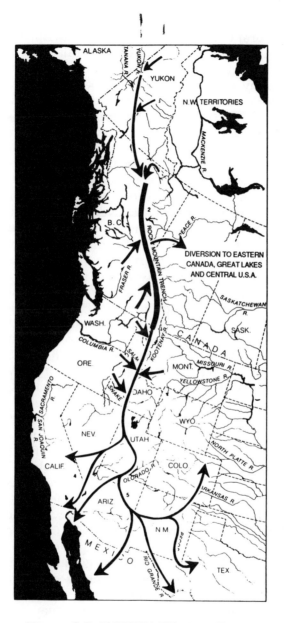

Figure 6.9: NAWAPA Western Region.
From "The Salinity of Rivers," by Arthur F. Pillsbury,
Copyright © July, 1981 by *Scientific American, Inc.*  All
rights reserved.

All natural waters, including those described as fresh, contain salts and other dissolved minerals. According to Pillsbury, a virgin stream emerging from a mountain watershed many contain as little as 50 parts per million (ppm) of total dissolved solids of which salt is the major constituent. Ocean water, by comparison, has about 35,000 ppm. As the major rivers flow through arid regions for great distances, as is the case with all the major western rivers, the concentration of salts rises steadily as a result of evaporation. Rivers would normally carry the salts and other minerals dissolved out of rock and soil to the oceans, but when the rivers are diverted for irrigation, the salt gets trapped in the soil, where it keeps accumulating.

The only effective means of flushing the salt out of the soil so it can continue its journey to the oceans is to irrigate the land with fresh water. The problem is that the fresh water that could accomplish the task is in the northern portions of Alaska and Canada. This vast quantity of fresh water (about 800 maf/year) is not presently being used by anyone. It just keeps flowing into the Arctic ocean where its fresh water status is terminated. A NAWAPA-scale project is what is required to bring fresh water to where it is critically needed if the long-term viability of irrigated agriculture in the U.S. is to be maintained.

One of the major disadvantages of the initial NAWAPA proposal was that it would involve constructing many hydroelectric power plants to serve as pumping stations along the way. Hydroelectric facilities are able to generate relatively low-cost electricity, but they cause extensive damage to natural ecosystems that are affected. The hydroelectric facilities in the initial proposal would be able to generate an electrical surplus of about 70,000 megawatts, which is equal to about 20 percent of present U.S. electrical production. However, 70,000 megawatts is equivalent to about 1.5 quads per year, which is only about 2.3 percent of the total amount of energy that is consumed in the U.S. annually. This being the case, it may make more sense to use the solar-generated electricity and/or hydrogen from the desert areas in the southwest U.S. and Mexico to power the NAWAPA pumping stations, and thereby avoid many of the environmental problems that are associated with hydroelectric dams.

The salt accumulation problems are in many respects like the energy and environmental problems. Few people are aware or concerned; very little, if anything, is being done; the problems are increasing exponentially; and the net result of inaction could be potentially catastrophic. It has been estimated by Pillsbury that within 15 years, over one-third of California's prime agricultural land (nearly 3 million acres), which now provides nearly one-half of the fruit and vegetables grown in the U.S., will be destroyed by salt accumulation. In addition, the water and salt problems are not just occurring in the West or Southwest. It has been estimated that 25 percent of all irrigated acreage in the U.S. is already developing serious salt accumulation problems. Moreover, like the energy and environmental problems, the obstacles to resolving the water problems are not technical -- they are political.

The NAWAPA proposal was initially expected to cost $200 billion (1965 dollars) and, like the large-scale production of solar-hydrogen gensets, the project would take 30 or 40 years to implement. In spite of the fact that the citizens in Canada, Mexico and the U.S. all need the water, nothing has been done to make the NAWAPA a reality. Indeed, the political interests in the Northwest U.S. have even lobbied not to have any further studies on NAWAPA undertaken. And, like the industrial transition to renewable resources, the longer the project is delayed, the more expensive it is ultimately going to be to implement. These are the types of long-term capital intensive macro-engineering projects that will not happen without strategic planning and cooperation between government and private industry. As of yet, no such cooperative planning effort exists or is anticipated. Indeed, in the present political environment, there is a generally negative attitude about anything that gives the impression of centralized planning. This conservative political belief makes the assumption that centralized Federal planning is incompatible with the notion of a free market.

Unfortunately, with this reasoning, the best plan is not to have a plan at all. There is no question about the destructive nature of the course this leads us to. The only question is: *Is there still time to change course?* There are many people who are either unaware of the problems we face, or who simply do not care. If these forces of apathy and ignorance prevail, the oblivion scenario for human civilization, and for the Earth's biological life-support systems will be inevitable.


*Conclusions*

This chapter has documented that there are more than ample renewable resources to provide for the energy needs of the human community. Indeed, there are many renewable resources and technologies that could be utilized for large-scale, multiple-quad hydrogen production. Photovoltaic, line-focus or point-focus Stirling dish genset systems, wind power, and ocean thermal (OTEC) systems are some of the most promising options that have thus far been developed. But there is a wide range of other renewable resources and technologies, including advanced photovoltaic cells; ocean currents and waves; nuclear fusion and biochemical energy systems that could promise to be cost-effective methods of producing hydrogen. However, one of the least expensive methods of hydrogen production may come from the nonphotosynthetic bacteria that live in the digestive tracts and wastes of humans and other animals.

In the final analysis, marketplace economics will determine which renewable energy technologies evolve. What needs to be understood, however, is that before the private sector industries commit themselves to a long-term and costly energy transition, there will have to be a clear mandate from both Federal and state officials. At present, the primary obstacle that is preventing such a mandate is the general lack of communication between scientists who tend to operate in highly specialized fields; between elected officials who formulate energy and natural resource policies; and most of all, between the news media and the voting public who have the power to influence their elected officials. One can only hope that this communications problem is resolved before irreparable damage is inflicted on the Earth's biological life-support systems.

Chapter 7

# CONCLUSIONS

A primary thesis of this book is that because of the exponential impact of the global environmental problems and the development of technology in general, humanity is on the threshold of both oblivion and a kind of technological utopia. Due to the complexity of the many environmental factors involved, it is not yet possible to predict accurately which outcome is going to evolve. But because virtually all of the significant global events are accelerating exponentially, there is little question that the transformation time is rapidly approaching. Roughly one-quarter of the human population is already seriously malnourished. That means in exponential terms, it is already approximately 11:58. The stunning advances in computers, artificial intelligence and molecular biology are continuing, but so are the many environmental problems that now threaten the survival of humanity itself.

From a comprehensive perspective, it seems clear that the existing stable-state environmental and economic systems are unsustainable. At the start of the Twentieth Century, there were roughly 1.5 billion people on the Earth, and virtually no automobiles. At present there are more than 5 billion people and hundreds of millions of automotive vehicles. If the present trends are able to continue, there will be more than six billion people and a perhaps a billion automotive vehicles by the year 2000.

Major food production systems are already in serious trouble due to the impending death of the oceans, global chemical contamination, salt accumulation, droughts and other climatic disruptions that are being intensified by global warming, stratospheric ozone holes, and deforestation. The warning signs are everywhere. It is hard to pick up a newspaper or magazine, or watch a television newscast that does not report on some serious global environmental or economic problem that is continuing to worsen. Given these obvious warning signs, it should be clear that the people on *Spaceship Earth* are like the passengers on the *Titanic*. There is no longer any question that we are going to hit the iceberg. The only questions are: Are there going to be any survivors, and if so, what is their existence going to be like?

*Solutions*

If civilization is to survive the many exponentially compounding energy and environmental problems that lie ahead, it will likely be dependent upon whether the following actions are implemented in time:

1. There needs to be a fundamental industrial transition from the use of fossil fuels and nuclear-fission facilities to renewable solar-hydrogen energy technologies and resources that will bring about a global "hydrogen economy."

2. There needs to be an immediate international effort to stop any further deforestation, coupled with a massive program of reforestation.

3. Strict international laws should be enacted to ban the production and use of toxic chemicals or products that are not biodegradable or are otherwise environmentally unacceptable. If a product cannot be recycled, it should not be made in the first place. New chemicals should not be viewed as innocent until proven guilty.

4. There needs to be a dramatic de-escalation of the arms race, particularly with respect to the U.S. and the Soviet Union. The armed forces in both countries should be more concerned about protecting the Earth's biosphere and endangered species rather than short-term commercial interests of private companies.

5. The U.S. should implement an industrial leadership council, similar to Japan's Ministry of International Trade & Industry (MITI) to establish and coordinate national industrial strategies. The priority should be to foster international cooperation among and between business and governmental leaders to solve the problems that threaten the Earth's biological life-support systems.

6. Lastly, and perhaps most importantly, there should be the widespread development of what are referred to as "controlled-environment" food production systems. Such "lifeboat" technologies could be the determining factor in whether or not civilization survives.

*Controlled-Environment Food Production Systems*

In order to avoid a collapse of global food production systems, it is critical to be prepared for the multiple environmental and related climatic dislocations that could eliminate the existing agricultural systems. One of the most important safeguards that could be implemented is what is referred to as controlled environment agriculture and aquaculture systems. Such automated indoor systems utilize artificial lighting to grow high-quality produce or seafood. Controlled-environment food production systems are like lifeboats because they can operate regardless of weather conditions; they require no conventional farm or fishing equipment; and no toxic pesticides or herbicides. Moreover, a one-acre controlled-environment facility in DeKalb, Illinois, called "Phytofarms of America, Inc.," has been able to grow as much produce as a 100-acre conventional outdoor farm, without any topsoil, with only one-tenth as much water, in about half the time. What controlled-environment agricultural systems *do* require is a reliable and economic source of energy.

The Phytofarm facility has been growing produce commercially for over 6 years. Noel Davis, a mechanical engineer who graduated from the Massachusetts Institute of Technology (MIT), initially developed the Phytofarm concept in the 1970's for General Mills Corporation.  Davis was able to purchase the rights to the Phytofarm process and technology in 1983, and has been operating the facility ever since.  The one-acre, $7 million Phytofarm facility currently produces about four tons of spinach and herbs weekly for supermarkets and restaurants in several states.  Similar success has been achieved with aquaculture systems for growing seafood.

The significance of indoor controlled environment systems is that they could quickly be expanded until enough fresh food could be produced to feed the U.S., and ultimately, the world, regardless of what happens to the ozone layer, the greenhouse gases, acid deposition, or topsoil erosion.  Moreover, such facilities could be located in the general vicinity where the food is to be consumed, thereby maximizing freshness and minimizing shipping.  This means that while the surface of the Earth might indeed become uninhabitable, human beings could continue to operate in artificial environments indefinitely.  While the thought of living in totally artificial environments may seem unpleasant, it is well to remember that most urban dwellers already spend most of their lives in such environments.  The Japanese are already planning to construct vast underground cities due to their serious land shortage problem.

Hopefully, it will be possible to avoid the complete destruction of the Earth's surface.  But it is now likely that a large number of the mammals now inhabiting the Earth's surface will be lost in the environmental and climatic dislocations that are probably inevitable.  Even in such a grim situation, however, indoor controlled-environment food production systems could allow human civilization to continue.  The basic premise is that as long as people can eat, they will not panic, and thereby survive.  The most important question is: *How many of these "lifeboat" systems can be developed before the more conventional food production systems fail?*  This underscores the fact that while there are many formidable long-term problems facing the human community, there are also many technical options available to deal with these problems.  It is only a question of priorities.

Figure 7.1: An external view of the Phytofarm controlled-environment agricultural facility in DeKalb, Illinois. (Courtesy of Phytofarms of America, Inc.)

Figure 7.2 An indoor view of the one-acre Phytofarm facility that has the same yield as a 100 acre conventional farm. (Reprinted with permission from James Kilkelly Photography © 1988, New York, N.Y.)

What is significant is that the primary obstacles to implementing such changes are not technical, or even economic; they are educational and political. Watching news reports about mass public demonstrations calling for democracy and freedom of the press in Europe and Asia is fascinating. What is troubling, however, is that such freedoms have been a part of U.S. society for generations, yet the U.S. political system, with its emphasis on money and superficial 30-second campaign slogans, is hardly an example of an ideal system.

Consider that the U.S. is now the largest debtor nation in the world; its student test scores are below those of all of the other major industrialized countries -- and even many Third World countries; toxic dump sites are in virtually every city and town (hundreds of thousands of them in total); and there is an epidemic of homelessness, drug abuse, violence and crime. Taken together, these problems seem insurmountable. That local, state and Federal governmental officials have been unable to reverse these serious trends in recent years does not inspire confidence that anything can or will be done. Understandably, those who can afford it move into communities with walls and guards at the gate, although such security systems will not be of much use if the supermarkets are empty.

These serious problems are compounded by the fact that the U.S. Congress is preoccupied with its own political scandals and relatively meaningless legislative efforts, such as passing a constitutional amendment to prevent a symbolic act of flag burning, rather than changing the policies that are causing the contamination and destruction of the Earth's ancient biological life-support systems. It brings to mind the phrase, "Rome burned while Nero fiddled," except in this case it is not just a city, or even a country, that is at stake. It is literally the survival of global ecosystems. One of the key conclusions of this book, however, is that humanity is as close to utopia as it is to oblivion. It is possible for everyone to be a survivor *if responsible changes are implemented in time.*

Given the power of the electronic and print media to focus the world's attention on key events, there is every reason to be hopeful that the necessary changes can be made in time. Consider the media attention that was given to three whales caught in the ice, the Exxon oil tanker spill in Alaska, or the cold fusion reaction reported in the press by chemists at the University of Utah. The cold fusion event was particularly interesting because most news reporters and members of Congress were apparently surprised to learn that someone may have discovered a process that could provide unlimited energy without producing significant amounts of radioactive waste, greenhouse gases, or acid rain. What is hard to understand is that the solar hydrogen energy systems discussed in Chapters 4 and 5 have the same advantages of the cold fusion nuclear process, without producing *any* radioactive or other toxic wastes. Moreover, the solar-hydrogen technologies are not unproven theoretical or laboratory curiosities that cannot even be discussed in terms of capital costs. Rather, they are well-documented technologies that are similar to an automobile in terms of cost estimates and manufacturing.

Although many people automatically make the assumption that the major multinational oil companies would never allow solar-hydrogen systems to be implemented, this book has explained why such oil companies will find it in their own economic self-interest to make such an energy transition. Unlike the traditional high-risk oil investments, renewable energy systems will provide a renewable rate of return with little risk, and without any of the problems associated with contaminating the natural environment. This, plus the fact that the Earth's solar energy resources are somewhat concentrated, will insure that the oil companies will be all too happy to evolve into pollution-free hydrogen companies.

Because the transition to a hydrogen energy system will solve many of the most serious energy and environmental problems, *it is a transition that both environmentalists and oil company executives will be able to support.* This is an extremely important consideration, because few other significant issues could attract the support of such divergent interest groups. This being the case, many people ask an obvious question: If the solar-hydrogen energy system has so many advantages, why don't the oil companies just start making hydrogen?

The question is understandable.

The answer is two-fold. First, one could accurately say that there is a fundamental information-gap between energy engineers. Engineers must increasingly specialize, and as a result, they can only be knowledgeable in relatively narrow fields. A good example is that virtually none of the solar dish engineering teams were aware of the hydrogen engineering community; and virtually none of the hydrogen engineering specialists were aware of the many solar dish engineering groups and their ability to produce base-load quantities of low-cost hydrogen. If most engineering specialists have not put the pieces together, it is not surprising that most elected officials have not seen "the big picture. "

In addition to the information gap, which is a direct result of the exponential knowledge explosion, there is also the fact that individual companies do not determine national energy policy. Even if all of the oil companies were ready to make the transition to hydrogen, they would have to make sure that the automobile and aircraft companies would follow a timetable in terms of modifying their vehicles to use the hydrogen fuel. Other infra-structure systems that involve pipeline companies, shipbuilding companies and hundreds of other manufacturers that make everything from stoves to water heaters must also be involved in the industrial transition timetable. What all this means is that a fundamental energy transition from oil to anything else is an undertaking that cannot be implemented without government leadership, coordination and support. In the same way that President Roosevelt called on industry leaders to help form the War Production Board to deal with World War II, President Bush now needs to set up a Hydrogen Production Board to coordinate a national and international industrial transition to renewable energy resources and technologies.

### The Bush Administration

In spite of the many environmental and scientific warning signs, President Bush has thus far demonstrated a "business-as-usual" attitude that clearly indicates he and his senior advisors are not aware of the solar-hydrogen energy technologies. His insistence in promoting nuclear power, offshore oil drilling and the continued oil exploration in the remaining wilderness areas, in spite of the Exxon-Alaskan oil spill, is especially worrisome. And although the Cold War appears to be coming to an end, President Bush is continuing to support the development of multi-billion dollar nuclear weapons such as the B-2 Stealth bomber and MX missile systems, rather than renewable energy systems. In this regard, President Bush is somewhat like the captain of the *Titanic* who refused to listen to the warnings of his engineering officers when they alerted him that there were icebergs up ahead.

To underscore this point, senior officials within the Bush Administration deliberately altered the testimony of Dr. James E. Hansen, director of NASA's Goddard Institute for Space Studies, in order to downplay the serious nature of the global warming greenhouse effect. The report was prepared for the Senate Subcommittee on Science, Technology and Space, and Hansen confirmed that the White House's Office of Management and Budget had altered key conclusions.

Hansen's original testimony had stated that computer projections of changes in climate caused by the human release of carbon dioxide and other greenhouse gases into the atmosphere predicted substantial increases in temperature, drought, severe storms, and other stresses affecting the Earth's biological systems. The text of Hansen's testimony, however, was changed by the officials in the budget office to suggest that the relative contribution of human and natural processes to changing climatic patterns "remains scientifically unknown." In fact, Hansen said, he and his colleagues at NASA are confident that the greenhouse gases are of human origin. Hansen responded to the alterations of his congressional testimony as follows:

> *"It distresses me that they put words in my mouth; they even put it in the first person...  I should be allowed to say what is my scientific position. There is no rationale by which OMB should be censoring scientific opinion. I can understand changing policy, but not science."* [1]

Hansen indicated he tried to negotiate with budget officials about the change in wording, but they refused to change. This is not surprising given the fact that White House officials have been urging a go-slow approach toward policies dealing with global warming. Senior Bush Administration officials even refused to support a global conference of nations to address the greenhouse problem. This was in contrast to the State Department and Environmental Protection Agency officials, who had been urging President Bush to take the lead in mobilizing the international community to address the problem. After a public outcry, President Bush reversed his position and agreed to let the U.S. host an international "workshop" (instead of conference) on the greenhouse issue, but one can hardly call such actions "leadership."

To his credit, President Bush has put forth new air quality proposals that would reduce acid deposition from fossil fuel power plants by 50 percent over a period of 10 years. However, such proposals must get through the Congress and the army of industry lobbyists who are opposed to such measures. President Bush also promoted the use of ethanol as an automotive fuel, but according to Dr. John Appleby, director of Texas A&M's Center for Electrochemical Systems and Hydrogen Research, ethanol, methanol and other alcohols derived from grain cannot contribute much toward solving the air-pollution or greenhouse problems. Although such fuels have only half the carbon content of gasoline, Appleby points out that it takes twice as much fuel to travel the same distance, which means the carbon content is about the same as gasoline for the same energy content[2]. Thus, making a transition to such "alternative fuels" is somewhat like rearranging the deck chairs on the *Titanic*.

This view is reinforced by a report released by the U.S. Office of Technology Assessment, which analyzes technical issues for Congress. The report, released in July of 1989, indicated that even if the most costly and stringent of current pollution control technologies were in widespread use by the year 2000, most major metropolitan cities would still be in violation of EPA air quality standards. Not surprisingly, the report found that automobiles and trucks are the largest single contributors to urban air pollution, along with oil refineries[3].

Although President Bush has acknowledged that the global environmental problems are serious, his policies to date have been woefully inadequate. If these policies are not changed during his Administration, the oblivion scenario may be inevitable. President Bush appears to be a friendly and likeable person, but thus far, his energy and environmental policies are directly contributing to a global systems collapse. The hour is already late. Given the exponential nature of the global environmental problems, eight critical years were lost in the Reagan Administration, which viewed environmental concerns as unnecessary governmental regulation. If the Bush Administration continues to follow the same course, the resulting death and destruction is going to be incalculable. It seems clear that major, rather than minor "band-aid" changes are called for. In acknowledgement of the seriousness of the global environmental crisis, *Time* magazine altered its normal policy of selecting a person of the year in 1989, and instead highlighted the fact that the Earth is in serious trouble. The following is excerpted from one of the main articles titled, "What on Earth Are We Doing?"

> *"Let there be no illusions. Taking effective action to halt the massive injury to the earth's environment will require a mobilization of political will, international cooperation and sacrifice unknown except in wartime...."*
>
> *"As man heads into the last decade of the 20th century, he finds himself at a crucial turning point: the actions of those now living will determine the future, and possibly the very survival of the species."*
>
> *"Now more than ever, the world needs leaders who can inspire their fellow citizens with a fiery sense of mission, not a nationalistic or military campaign but a universal crusade to save the planet. Unless mankind embraces that cause totally, and without delay, it may have no alternative to the bang of nuclear holocaust or the whimper of slow extinction."* [4]

While many people understandably mistrust governmental regulatory agencies, the fact remains that these are
the only institutions that can force private companies to
comply with the regulations that are deemed to be in the
public interest. The problem is that the U.S. Government
is in a state of paralysis, and the governments of most
other countries are worse off than the U.S. with respect to
having -- much less enforcing -- environmental guidelines.
The sad reality is that at a time of the greatest peril in
history, there seems to be no one at the helm.

One might ask, why don't the American students and
citizens rise to the occasion, and demand fundamental
changes. Given the extensive media coverage of the many
environmental problems, there is certainly no question
that most people are aware there are life-threatening problems. Polls reinforce this view.

The problem seems to be that the President, the
members of Congress, the general public, and the media
which generally informs them, is not aware of a specific
plan that can resolve the multiple and incredibly complex
global problems. After all, being aware of the problems is
one thing. Being aware of the solutions is quite another.
Even those individuals who take the time to be informed
are likely to throw up their hands and say, *If the "experts"
don't have a plan, how is the average citizen supposed to
come up with one?* But as this book has documented,
there are indeed solutions to many of the most serious energy and environmental problems. Humanity is as close to
utopia as it is to oblivion, and whether one or the other
evolves will ultimately depend on *citizen action.*

Elected officials generally follow public opinion polls
with great care, for their political survival depends upon
serving the interests of their constituents. What this
means, however, is that elected officials are by definition
followers and not leaders. The leaders are, therefore, the
majority of voters, and until they demand that their
governmental representatives make a transition to renewable resources, it is highly unlikely that such a fundamental change will happen. Fortunately, there is already legislation that has been introduced in the U.S.
Congress that could begin to reestablish a national energy
strategy around a hydrogen energy system.

*Senate Bill 639: A Bill For All Interests*

While a Hydrogen Production Board is not presently being considered by anyone in the U.S. Congress, more moderate hydrogen legislation has been co-sponsored in the Senate for several years by Senator Spark Matsunaga (D-Hawaii) and Senator Dan Evans (R-Washington). It is worth noting that Senator Evans was one of the few professional engineers to ever serve in the U.S. Senate. Companion legislation to Senate Bill 639 (HR 2793) has been introduced in the U.S. House of Representatives by Congressman George Brown (D-California), who is the only physicist currently serving in either the House or Senate. It is significant that those with technical training are sponsoring the hydrogen development legislation. One of the main provisions of Senate bill 639 calls for the Secretary of Energy to prepare and implement a comprehensive five-year plan and program to:

> *"...accelerate research and development activities leading to the realization of a domestic capability to produce, distribute, and use hydrogen economically within the shortest time practicable."*

The proposed hydrogen legislation also calls for the development of a "Hydrogen Technical Advisory Panel" to advise the Energy Secretary. Appropriations for Title 1 of the hydrogen legislation call for $55 million to be spent over five years. Title II, which deals with the development of hydrogen-fueled aircraft research and development, calls for an additional $100 million to be spent over the same five-year period. This is in contrast to current Federal spending for hydrogen fuel research that amounts to only about $2.4 million annually. What needs to be put into perspective is that in the same five year period, hundreds of billions of dollars are going to be spent on keeping U.S. troops stationed in Europe and elsewhere, as well as designing, building, and maintaining ever more sophisticated weapons of mass destruction. This only underscores the fact that it is not a question of money, but of national political priorities.

# Congressional Record

United States of America

PROCEEDINGS AND DEBATES OF THE $101^{st}$ CONGRESS, FIRST SESSION

| Vol. 135 | WASHINGTON, FRIDAY, MARCH 17, 1989 | No. 31 |

## Senate

### HYDROGEN LEGISLATION: A BILL FOR ALL INTERESTS

Mr. MATSUNAGA. Mr. President, I rise to introduce, together with my senior colleague from Hawaii, Senator INOUYE, the junior Senator from Colorado, Senator WIRTH, the senior Senator from Rhode Island, Senator PELL, and the junior Senator from Connecticut, Senator LIEBERMAN, legislation which advances a multitude of causes.

Seldom can a single bill be said to address the gamut of legislative issues, both environmental and economic, facing a new Congress and a new administration. After all, what common thread exists between global climatic change and those activities spawning greenhouse gases, on the one hand, and the U.S. trade deficit and American competitiveness in world markets, on the other?

Energy is the thread, Mr. President, and a national program for hydrogen research and development is the measure that touches all the aforementioned bases.

The form of energy we use is at the heart of virtually all environmental issues. The location of our energy sources is at the crux of our trade imbalance. And in no area of economic endeavor is overseas competition more keenly felt than in the quest for new energy technologies.

For all these reasons I am once more introducing the National Hydrogen Research and Development Program Act, legislation I first offered in the 97th Congress and have urged ever since. Mine has not been a lone voice on this subject, Mr. President. Throughout his career in this Chamber former Senator Dan Evans cosponsored and strongly advocated my hydrogen legislation. In this connection, it is significant that Senator Evans has been the only professional engineer to serve in the Senate in recent memory, save for Mike Mansfield, who was a mining engineer, and the late Stewart Symington, who was self-taught in mechanical and electrical engineering. If they were still with us, I am certain that both would be cosponsors, Mr. President. Moreover, the principal sponsor of companion legislation in the House, Representative GEORGE BROWN of California, is one of the very few scientists serving in that body.

Mr. President, hydrogen is one of the most abundant elements in the universe, with water, a primary source of hydrogen, covering three-fourth of the Earth. Indeed, hydrogen plays a role in such varied, everyday products as peanut butter, vitamin C, and aspirin, not to mention such larger products as clear plate glass windows.

As a transportation fuel, hydrogen's environmental benefits are particularly apparent, as was evident by its inclusion in the national energy policy legislation offered by Senator WIRTH in the last Congress to address the concerns over global warming, acid rain and the greenhouse effect and introduced again this year. Moreover, hydrogen can be transported more efficiently and at less cost than electricity over long distances.

While hydrogen has definite environmental advantages over fossil fuels, because the product of hydrogen combustion with air is essentially water vapor, it also offers benefits in the utilization of numerous energy alternatives—ranging from coal and natural gas, to nuclear as well as to solar and the renewables. Injected into declining natural gas fields, hydrogen can serve as an enhancer, stretching out the life of dwindling supplies.

For those concerned with the interests of the coal industry, hydrogen also figures in an attractive scenario. If coal-gased reactors were to be built at the seashore, they could eject carbon dioxide into the sea instead of into the air, and transmit energy in the form of hydrogen from coal. It is claimed that this could give us perhaps another half century of coal availability without adding anything to the world greenhouse effect.

For those interested in advancing nuclear power, hydrogen can be seen as a vehicle for hurdling the safety barrier. Because energy is cheap to transport long distances with hydrogen as the storage medium and after 300 to 400 miles, increasingly cheaper

Figure 7.3: Hydrogen Legislation.

Ultimately, it is literally a question of continuing to make massive investments in the technologies of death, or changing priorities to invest in the technologies of life. It would seem that any reasonable person could understand the importance of protecting the Earth's biological life support system. Yet, the environmental habitats of the Earth are being lost at a terrifying rate. As an old saying goes, "*We have found the enemy, and it is us.*"

Two of the more important considerations are that it is critical to take action while there is still time to make a difference, and that major -- and not minor -- changes are called for. Moreover, if the leadership in changing national priorities is not provided by elected officials within the state and national governments, then it must come from the public at large. However, before such a public initiative is going to occur, the public needs to be made aware that there are viable solutions to the most serious problems that we collectively face, and that the essence of the solutions involves making an industrial transition to the types of renewable energy resources and technologies that have been documented in this book.

When the industrial transition to solar-hydrogen technologies and resources gets the kind of intense media exposure that the alleged cold fusion experiments received, the American public will demand that the transition begin as soon as possible, and that it be implemented with war time speed. It is hoped that this book will help to advance this critical energy and industrial transition before the global environmental and economic disruptions make such a transition impossible.

## Government Regulation

Although there are a great many people who sincerely believe that government regulation is the problem and a less regulated economy is the solution, it is well to remember that many of the most serious environmental problems have been the result of a lack of governmental regulation. Indeed, most regulations are only enacted after serious problems have already occurred.

History has repeatedly shown that a true free enter-
prise system requires a reasonable amount of government
regulation if it is to remain free.  There is no better example than what happened with John D. Rockefeller.  When
he started out in the oil business, the market was a highly
competitive one.  But over time, Rockefeller became more
and more successful (some would say ruthless) than his
competitors, and he was eventually able to buy them out,
and in so doing, he created a virtual monopoly that effectively ended the free market forces in the oil business.
Thus it would seem that an unregulated free market contains the seeds of its own destruction.

While governmental agencies are far from perfect, they
are the only institutions that are responsible for looking
after the health and welfare of the general public.  As a
result, a more balanced and realistic approach would involve establishing a partnership between the private sector
and the governmental regulatory agencies.  Using the
analogy of a football game, private companies should be
like the players in the game while the government officials
are the referees, who establish and enforce the rules.  As
long as everyone competes fairly under the established
rules, there is no reason for the governmental agencies to
interfere with the corporate players.

The importance of governmental officials is that they
are supposed to be somewhat objective and represent all of
the people.  This is in sharp contrast to individuals working for private companies who are expected to look out for
their company's self-interest.  It is not an "either-or" question of whether an unregulated economy is inherently bad
or good.  It is a matter of acknowledging how a profit-ori-
ented free-market system works.  Private companies cannot be expected to take on social responsibilities which are
not in their own short-term economic self-interest.  While
most companies seek to cooperate with the government,
few -- if any -- want to assume the responsibilities of the
government to protect the public welfare.  It is simply not
their function.

Acknowledging the inherent nature of companies to
focus on short-term profitability at the expense of long-
term planning, research and development, Japan has developed a highly successful Ministry of International Trade

& Industry (MITI) that has been responsible for co-ordinating Japan's long-term industrial strategy.

In sharp contrast, U.S. companies essentially make business decisions in isolation. Corporate executives rarely have input from governmental officials on important corporate business decisions, and most companies would never think of consulting with their competitors on such matters, which in many cases, may actually be illegal in the U.S. Japan, on the other hand, has been able to evolve into an economic and technological superpower by having its private companies work in partnership with MITI. Such strategic planning would seem to be necessary if the U.S. is to remain competitive in the 1990's and beyond. The magnitude and severity of the problems confronting the human community underscores the need for a balanced approach to optimizing industrial and environmental policy decisions, where cooperation is valued as well as competition.

### Final Comments

Given the serious nature of the global environmental problems, it is time for elected officials to shift from a policy of attempting to *manage* the problems, and to take the fundamental steps necessary to *solve* them.

One of the more significant first steps that needs to be taken is for Federal and state governments to provide the leadership and guidelines that will initiate a national and international transition to renewable energy resources. In the same way that the U.S. Environmental Protection Agency and the California Legislature forced the major automobile companies to put catalytic converters on automobiles, they could mandate that after a certain year, no automotive vehicle sold could be allowed to emit *any* hydrocarbon exhausts. That would force the automotive companies to develop either hydrogen-fueled or electric-powered vehicles. On a Federal level, the U.S. government needs to establish a national energy policy that will allow the industrial transition to renewable resources to occur as rapidly as possible.

Given that a large number of contaminants have already been released into the global environmental systems, it is now likely that a number of ecological system collapses will occur. This will drastically reduce the existing human population levels, and most of the Earth's remaining ancient wilderness areas. But human adaptability being what it is, it is reasonable to expect that with a successful energy transition to renewable resources and the rapid expansion of controlled-environment agricultural systems, human civilization will be able to continue. This, in turn, will insure that the computer and biotechnology advances will continue. Given that medical researchers in molecular biology have already begun to regenerate organs and other biological tissue in mammals, it is now only a question of time before living organisms make a biological transition to renewable resources. With the exponential advances in molecular biology and computer science, it would be ironic indeed if the civilization that allowed such truly astonishing developments to occur disintegrated because it also allowed its own biological life-support systems to be irreparably contaminated and/or destroyed.

One thing is clear: humanity now stands at the threshold of the end of life as we know it, and given the exponential nature of the events now unfolding, the oblivion and/or utopia scenario will occur sooner than most people expect. This underscores the sense of urgency for the public to demand decisive action by their elected representatives. If there are those who say they have a better solution to the global environmental problems than making an industrial transition to renewable solar-hydrogen energy technologies and resources, let them come forth and demonstrate a viable alternative. If not, then let's get on with the job that needs to be done. We have waited long enough, for the time to stand and be counted is rapidly slipping away.

\*          \*          \*

# REFERENCES

## CHAPTER 1:
## UTOPIA OR OBLIVION

1.  J. O'M. Bockris, "Hydrogen Economy," SCIENCE (American Association for the Advancement of Science), Vol. 176, No. 4041, p. 1323, June 23, 1972.

2.  Derek P. Gregory (Institute of Gas Technology), "The Hydrogen Economy," SCIENTIFIC AMERICAN, Vol. 228, No. 1, pp. 13-21, January 1973.

3.  T. Nejat Veziroglu and A. N. Protsenko, *Hydrogen Energy Progress VII: Reviewing the Progress in Hydrogen Energy*, PERGAMON PRESS (New York, New York), October 1988.

4.  J. Pangborn, M. Scott and J. Sharer, "Technical Prospects for Commercial and Residential Distribution and Utilization of Hydrogen," INTERNATIONAL JOURNAL OF HYDROGEN ENERGY (International Association for Hydrogen Energy), Vol. 2, pp. 431-445, 1977.

5.  Tom Wicker, "Chilling Thoughts of Summer," THE NEW YORK TIMES release, published in the ARIZONA REPUBLIC (Phoenix, Arizona), p. A13, January 17, 1989.

6.  Mary Wayne, et al., "Acid Rain: Clarifying the Scientific Unknowns," ELECTRIC POWER RESEARCH INSTITUTE (EPRI) JOURNAL, Vol. 8, No. 9, p. 8, November 1983.

7.  Volker A. Mohnen, "The Challenge of Acid Rain," SCIENTIFIC AMERICAN, Vol. 259, No. 2, pp. 30-38, August 1988.

8.  "Caustic fog poses worse threat to health, ecology than acid rain," LOS ANGLES TIMES release, published in THE ARIZONA REPUBLIC, p. A5, November 12, 1982.

9. "Acid-fog hazard is feared worse than announced," UNITED PRESS INTERNATIONAL release, published in THE ARIZONA REPUBLIC, p. A5, November 15, 1982.

10. Robert W. Shaw (chief of chemical diagnostics & surface science at the U.S. Army research office in Research Triangle Park, North Carolina), "Air Pollution by Particle," SCIENTIFIC AMERICAN, Vol. 257, No. 2, pp. 96-103, August 1987.

11. Richard A. Houghton and George M. Woodwell (Woods Hole Research Center, Woods Hole, Massachusetts), "Global Climatic Change," SCIENTIFIC AMERICAN, Vol. 260, No. 4, p. 36, April 1989.

12. Philip Shabecoff, "Greenhouse effect here, expert fears," NEW YORK TIMES release, published in THE ARIZONA REPUBLIC, p. A1, June 24, 1988.

13. Syukuro Manabe and Richard T. Wetherald (National Oceanic and Atmospheric Adminstration's Geophysical Fluid Dynamics Laboratory, Princeton, University), "Reduction in Summer Soil Wetness Induced by an Increase in Atmospheric Carbon Dioxide," SCIENCE, Vol. 232, No. 4750, pp. 626-628, May 2, 1986.

14. "Tampering with the Global Thermostat," COMPRESSED AIR MAGAZINE, Vol. 91, No. 10, pp. 16-22, October 1986.

15. Irwin W. Sherman and Vilia G. Sherman, *Biology: A Human Approach* (Third Edition), pp. 599-600, OXFORD UNIVERSITY PRESS (New York, New York), 1983.

16. "Tampering with the Global Thermostat," op. cit., p. 21.

17. James E. Lovelock, *Gaia: A New Look at Life on Earth*, OXFORD UNIVERSITY PRESS (New York, New York), 1979.

18. James E. Lovelock, *The Ages of Gaia: A Biography of Our Living Earth*, W.W. NORTON & COMPANY, INC. (New York, New York), 1988.

19. Houghton, op. cit., p. 39.

20. James F. Kasting, Owen B. Toon and James B. Pollack, "How Climate Evolved on the Terrestrial Planets," SCIENTIFIC AMERICAN, Vol. 258, No. 2, pp. 90-97, February 1988.

21. Gary Taubes, "Made in the Shade? No Way," DISCOVER, Vol. 8, No. 8, p. 65, August 1987.

22. "Diminishing Ozone called Major Threat," THE ARIZONA REPUBLIC, p. A1, March 10, 1987.

23. "The Heat Is On," TIME MAGAZINE, Vol. 130, No. 16, p. 62, October 19, 1987.

24. Peter Aleshire, "Despite steps to halt it, pollution will still reduce ozone," THE ARIZONA REPUBLIC, p. AA2, April 10, 1988.

25. Karen E. Bettacchi, et al., "How Man Pollutes His World," NATIONAL GEOGRAPHIC MAGAZINE (Washington, D.C.), 1970.

26. Ibid.

27. Linda Harrar, "The Hole in the Sky," NOVA, a Public Broadcast System (PBS) Science Documentary produced by the WGBH Educational Foundation, (Box 322, Boston, Massachusetts), 1987.

28. "Diminishing ozone called major threat: Skin-tumor death rate forecast to skyrocket," THE ARIZONA REPUBLIC, p. A1, March 10, 1987.

29. Shabecoff, op. cit., p. A1.

30. Richard J. Wurtman (Massachusetts Institute of Technology), "The Effects of Light on the Human Body," SCIENTIFIC AMERICAN, Vol. 233, No. 1, pp. 68-77, July 1975.

31. Robert M. Neer, T. R. A. Davis, A. Walcott, et al., "Stimulation by Artificial Lighting of Calcium Absorption in Elderly Human Subjects," NATURE, Vol. 229, No. 5282, pp. 225-226, January 22, 1971.

32. A.J. Lewy, et al., "Light Suppresses Melatonin in Humans," SCIENCE, Vol. 210, No. 4475, pp. 1267-1269, December 1980.

33.    "Ozone 'worse than acid rain' in short term," BOSTON GLOBE release, published in THE ARIZONA REPUBLIC, p. A4, December 26, 1985.

34.    "Nation's water tainted by leaks in gasoline tanks," NEW YORK TIMES release, published in THE ARIZONA REPUBLIC, p. A1, November 30, 1983.

35.    "1 million Americans likely to get cancer in '89, report says," UNITED PRESS INTERNATIONAL release, published in THE ARIZONA REPUBLIC, p. A10, February 25, 1989.

36.    Paul Raeburn, "Aid, not drought, caused African Famine," ASSOCIATED PRESS release, published in THE ARIZONA REPUBLIC, p. A3, November 20, 1985.

37.    Jeff Nesmith, "Evidence shows Earth heating up," COX NEWS SERVICE, published in the MESA TRIBUNE (Mesa, Arizona), p. A20, November 27, 1986.

38.    Houghton, op. cit., p. 36.

39.    Ibid., p. 44.

40.    William D. Ruckelshaus, "Toward A Sustainable World," SCIENTIFIC AMERICAN, Vol. 261, No. 3, pp. 166-174, September 1989.

## CHAPTER 2:
## EXPONENTIAL ICEBERGS

1.    Irwin W. Sherman and Vilia G. Sherman (University of California at Riverside), Biology: A Human Approach (Third Edition), OXFORD UNIVERSITY PRESS, p. 614, 1983.

2.    Albert A. Bartlett, "Forgotten Fundamentals of the Energy Crisis," AMERICAN JOURNAL OF PHYSICS (American Association of Physics Teachers), Vol. 46, No. 9, pp. 876-888, September 1978.

3.    Ibid., p.880.

4.    Sherman, et al., p. 614.

5.    Carl Sagan, *Cosmos*, RANDOM HOUSE (New York, New York), p. 281, May 1980.

6.    Alan Toffler, *Future Shock*, RANDOM HOUSE (New York, New York), p. 26, 1970.

## CHAPTER 3:
## CONVENTIONAL ENERGY
## CONSIDERATIONS

1.    Harrison Brown, "Energy In Our Future," reprinted by John C. Bailor, Jr., et al, *Chemistry*, ACADEMIC PRESS, A Subsidiary of Harcourt Brace Jovanovich (New York, New York), p. 843, 1978.

2.    Edward H. Thorndike, *Energy and Environment: A Primer for Scientists and Engineers*, ADDISON-WESLEY PUBLISHING COMPANY (Reading, Massachusetts), p. ix, August 1978.

3.    M. A. K. Lodhi and R. W. Mires, "How Safe is the Storage of Liquid Hydrogen," INTERNATIONAL JOURNAL OF HYDROGEN ENERGY, Vol. 14, No. 1, p. 36, 1989.

4.    "International Energy Annual, 1986," ENERGY INFORMATION ADMINISTRATION, U.S. DEPARTMENT OF ENERGY (DOE) (Washington, D.C.), p. 1, October 1987.

5.    Ibid., p. 21.

6.    Donald F. Othmer, "Energy -- Fluid Fuels from Solids," MECHANICAL ENGINEERING (American Society of Mechnical Engineers), Vol. 99, No. 11, pp. 29-35, November 1977.

7.    International Energy Annual, op. cit., p. 15.

8.    Albert A. Bartlett (Department of Physics, University of Colorado), "Forgotten Fundamentals of the Energy Crisis," AMERICAN JOURNAL OF PHYSICS, Vol. 46, No. 9, p. 881, September 1978.

9. Charles D. Masters, "World Petroleum Resources," Open-File Report No. 85-248, U.S. DEPARTMENT OF THE INTERIOR GEOLOGICAL SURVEY (USGS), September 1985.

10. Andrew R. Flower, "World Oil Production," SCIENTIFIC AMERICAN, Vol. 238, No. 3, pp. 42-49, March 1978.

11. "Fuel discovery outlook in U.S. dim, study says," NEW YORK TIMES release, reprinted in THE ARIZONA REPUBLIC, p. 1 & A17, Sunday, April 12, 1981.

12. Allen E. Murray (Mobil Oil Corporation), "The Impending Energy Crisis," NEWSWEEK MAGAZINE, Vol. 105, No. 23, p. 16, June 10, 1985.

13. M. K. Hubbert, "Energy Resources: A Report to the Committee on Natural Resources," NATIONAL ACADEMY OF SCIENCES-- NATIONAL RESEARCH COUNCIL, p. 1, Publication 1000-D, Washington D.C., 1962.

14. International Energy Annual, op. cit., p.25.

15. S. Harwood, et al., "The Cost of Turning It Off," ENVIRONMENT, Vol. 18, No. 10, p. 17, December 1978.

16. Richard Severo, "Too hot to handle,"(cover story) NEW YORK TIMES MAGAZINE, p. 15, April 10, 1977.

17. Ibid., pp. 15-22.

18. Mark Trahant, "Idaho is Winner in Nuclear 'War,' " THE ARIZONA REPUBLIC, p. A8, February 11, 1989.

19. Mark Trahant, "Storage Plan Capsulizes Nuclear Worries," THE ARIZONA REPUBLIC, pp. A1-A8, February 11, 1989.

20. "Cancer Leaps Blamed on Chernobyl Fallout," ASSOCIATED PRESS release, reprinted in THE ARIZONA REPUBLIC, p. B7, February 19, 1989.

21. Richard Lipkin, "A Safer Breed of Reactor in Sight" (Integral Fast Reactor), INSIGHT, pp. 52-53, January 23, 1989.

22. John Horgan, "Fusion's Future," SCIENTIFIC AMERICAN, Vol. 260, No. 2, pp. 25-28, February 1989.

23. "Fusion claims from Utah fail to gain backing," NEW YORK TIMES release, published in THE ARIZONA REPUBLIC, p. A11, April 29, 1989.

24. Hearings, 95th Congress, "Nuclear Power Costs," HOUSE OF REPRESENTATIVES, Report No. 95-1090, pp. 11-12, Committee on Government Operations, April 26, 1978.

25. Barry Commoner, *The Closing Circle*, BANTAM BOOKS (New York, New York), p. 61, 1972.

26. Kenneth and David Brower, "Miracle Earth," OMNI, Vol. 1, No. 1, pp. 16-18, October 1978.

## CHAPTER 4: HYDROGEN

1. Carl Sagan, *Cosmos*, RANDOM HOUSE (New York, New York), p. 233, 1980.

2. Daniel L. Alkon, "Memory Storage and Neural Systems," SCIENTIFIC AMERICAN, Vol. 261, No. 1, pp. 44-45, July 1989.

3. Carl R. Woese, "Archaebacteria," SCIENTIFIC AMERICAN, Vol. 244, No. 6, p. 100, June 1981.

4. C. Marchetti, "From the Primeval Soup to World Government--An Essay on Comparative Evolution," INTERNATIONAL JOURNAL OF HYDROGEN ENERGY, Vol. 2, pp. 1-5, 1977.

5. S. C. Huang, C. K. Secor, R. Ascione and R. M. Zweig, "Hydrogen Production by Non-Photosynthetic Bacteria," INTERNATIONAL JOURNAL OF HYDROGEN ENERGY, Vol. 10, No. 4, pp. 227-231, 1985.

6. Michael D. Levitt, "Production and Excretion of Hydrogen Gas in Man," NEW ENGLAND JOURNAL OF MEDICINE (Massachusetts Medical Society), Vol. 281, No. 3, pp. 122-127, July 17, 1969.

7.    M. Bigard, P. Gaucher and C. Lassalle, "Fatal colonic explosion during colonoscopic polypectomy," GASTROENTEROLOGY (American Gastroenterological Association), Vol. 77, No. 6, pp. 1307-1310, December 1979.

8.    Peter Hoffmann, *The Forever Fuel: The Story of Hydrogen*, pp. 204-205, WESTVIEW PRESS (Boulder, Colorado), August 1981.

9.    Irwin W. Sherman and Vilia G. Sherman (University of California at Riverside), *Biology: A Human Approach* (Third Edition), pp. 599-600, OXFORD UNIVERSITY PRESS (New York, New York), 1983.

10.    Abraham Lavi and Clarence Zener, "Plumbing the Ocean's Depths: A New Source of Power," IEEE SPECTRUM (Institute of Electrical & Electronics Engineers), Vol. 10, No. 10, pp. 22-27, October 1973.

11.    J. O'M. Bockris, "Hydrogen Economy," SCIENCE, Vol. 176, No. 4041, p. 1323, June 23, 1972.

12.    Derek P. Gregory (Institute of Gas Technology), "The Hydrogen Economy," SCIENTIFIC AMERICAN, Vol. 228, No. 1, pp. 13-21, January 1973.

13.    Medard Gabel, *Energy, Earth and Everyone*, STRAIGHT ARROW BOOKS (San Francisco, California), p. 141, 1975.

14.    "The Energy Crisis: One Solution," a film documentary produced by the College of Engineering, BRIGHAM YOUNG UNIVERSITY (Utah), in cooperation with Billings Energy Corporation (Provo, Utah), 1976.

15.    Walter Peschka (German Aerospace Research Establishment), "The Status of Handling and Storage Techniques for Liquid Hydrogen in Motor Vehicles," INTERNATIONAL JOURNAL OF HYDROGEN ENERGY, Pergamon Press, Vol. 12, No. 11, pp. 753-764, 1987.

16.    G. Daniel Brewer (Lockheed Aircraft Corporation), "Hydrogen Usage in Air Transportation," INTERNATIONAL JOURNAL OF HYDROGEN ENERGY, Vol. 3, No. 2, pp. 217-229, 1978.

17. W. T. Mikolowsky and L.W. Noggle (The Rand Corporation), "The Potential of Liquid Hydrogen as a Military Aircraft Fuel," INTERNATIONAL JOURNAL OF HYDROGEN ENERGY, Vol. 3, No. 4, pp. 449-460, 1978.

18. G. Daniel Brewer, op. cit.

19. Walter F. Stewart, "A Liquid Hydrogen-Fueled Buick," LOS ALAMOS NATIONAL LABORATORY (Los Alamos, New Mexico), Report No. LA-8605-MS, p. 7, November 1980.

20. "Liquid Hydrogen as a Vehicular Fuel," a report published by DEUTSCHE FORSCHUNGS-U. VERSUCHSANSTALT F. LUFT-U. RAUMFAHRT (i.e., the German Aerospace Research Establishment [DFVLR], Stuttgart, Germany), received in 1982.

21. Hoffmann, op. cit., p. 128.

22. Stewart, op. cit., p. 9.

23. R. D. Quillian, et al., (U.S. Army Fuels & Lubricants Research Laboratory, San Antonio, Texas), *Hydrogen Energy*, Part B, PERGAMON PRESS (New York, New York), p. 1025, March 1975.

24. Willis M. Hawkins and G. D. Brewer (Lockheed Aircraft Corporation), "Alternative Fuels Make Better Airplanes: Let's Demonstrate Now," ASTRONAUTICS & AERONAUTICS (American Institute of Aeronautics and Astronautics), Vol. 17, No. 9, pp. 42-46, September 1979.

25. Ibid., p. 43.

26. Hoffmann, op. cit., p. 211.

27. Paul M. Ordin (NASA Lewis Research Center), "Review of Hydrogen Accidents and Incidents in NASA Operations," 9TH INTERSOCIETY ENERGY CONVERSION ENGINEERING CONFERENCE PROCEEDINGS, (held in San Francisco, California), Technical Paper No. 749036, (American Society of Mechanical Engineers) pp. 442-453, August 1974.

28. B. C. Dunnam, "Air Force Experience in the Use of Liquid Hydrogen as an Aircraft Fuel," Proceedings of the FIRST WORLD HYDROGEN ENERGY CONFERENCE PROCEEDINGS, University of Miami (Miami, Florida), pp. 991-1010, March, 1974.

29. A. M. Momenthy (The Boeing Commercial Airplane Company), "Fuel Subsystems for Liquid Hydrogen Aircraft: R & D Requirements," INTERNATIONAL JOURNAL OF HYDROGEN ENERGY, Vol. 2, pp.155-162, 1977.

30. Gerard K. O'Neill, "The colonization of space," PHYSICS TODAY, Vol. 27, No. 9, pp. 32-40, September 1974.

31. Gerard K. O'Neill, *The High Frontier: Human Colonies in Space*, BANTAM BOOKS (New York, New York), 1976.

32. Carl Sagan, *Cosmos*, RANDOM HOUSE, (New York, New York), p. 206, May 1980.

33. Ibid., p. 204.

34. Howard P. Harrenstien (College of Engineering, University of Miami), "Hydrogen to Burn," OCEANUS (Woods Hole Oceanographic Institution),Vol. 17, pp. 28-29, Summer 1974.

35. Derek P. Gregory, et al. (Institute of Gas Technology), "The Economics of Hydrogen," CHEMTECH (American Chemical Society), Vol. 11, No. 7, pp. 432-440, July 1981.

36. C. R. Baker (Linde Division, Union Carbide Corporation), "Efficiency and Economics of Large Scale Hydrogen Liquefaction," SOCIETY OF AUTOMOTIVE ENGINEERS, Technical Paper No. 751094), delivered to the National Aerospace Engineering and Manufacturing Conference (held in Culver City, California), SAE Reference Vol. 84, p. 132, November 17-20, 1975.

37. C. A. Rohrmann and J. Greenborg (Battelle-Pacific Northwest Laboratories), "Large Scale Hydrogen Production Utilizing Carbon in Renewable Resources," INTERNATIONAL JOURNAL OF HYDROGEN ENERGY, Vol. 2, pp. 31-40, 1977.

38. Huang, op. cit., pp. 227-231.

39.     Willis M. Hawkins (Lockheed Aircraft Corporation), Letter to the Editor, INTERNATIONAL JOURNAL OF HYDROGEN ENERGY, Vol. 7, No. 1, p. 98, 1982.

40.     Refer to the INTERNATIONAL ASSOCIATION FOR HYDROGEN ENERGY (IAHE), P.O. Box 248266, Coral Gables, Florida, 33124. T. Nejat Veziroglu, Editor in Chief.

41.     An editorial by Dr. T. Nejat Veziroglu (College of Engineering, University of Miami), "Hydrogen Energy System: Next Action," INTERNATIONAL JOURNAL OF HYDROGEN ENERGY, Vol. 11, No. 1, pp. 1-2, 1986.

42.     Carl Sagan, "Cosmos," PUBLIC BROADCASTING SYSTEM (PBS): Excerpted from the television documentary that was initially aired in 1982.

## CHAPTER 5:
## SOLAR TECHNOLOGIES

1.      Wilson Clark, *Energy for Survival*, ANCHOR PRESS / DOUBLEDAY (New York, New York), p. 513, 1974.

2.      Ben Kocivar, "Tornado Turbine," POPULAR SCIENCE, Vol. 210, No. 1, p. 1 (Cover) and pp. 78-80, January 1977.

3.      William E. Heronemus (College of Engineering, University of Massachusetts at Amherst), "Using Two Renewables," OCEANUS (Woods Hole Oceanographic Institution), Vol. 17, p. 24, Summer 1974.

4.      Clarence Zener (Carnegie Mellon University), "Solar Sea Power," PHYSICS TODAY (American Institute of Physics), Vol. 26, No. 1, pp. 48-53, January 1973.

5.      W. H. Avery, D. Richards and G. L. Dugger (Applied Physics Laboratory, Johns Hopkins University), "Hydrogen Generation by OTEC Electrolysis, and Economical Energy Transfer to World Markets Via Ammonia and Methanol," INTERNATIONAL JOURNAL OF HYDROGEN ENERGY, Vol. 10, No. 11, September 1985.

6.    Donald F. Othmer and Oswald A. Roels, "Power, Fresh Water, and Food from Cold, Deep Sea Water," SCIENCE, Vol. 182, No. 4108, pp. 121-125, October 12, 1973.

7.    Donald F. Othmer, "Power, Fresh Water, and Food from the Sea," MECHANICAL ENGINEERING, Vol. 98, No. 9, pp. 27-34, September 1976.

8.    Heronemus, op. cit., p. 27.

9.    Robert L. Pons (Ford Aerospace & Communications Corporation, Newport Beach, California), "Optimization of a Point-Focusing Distributed Receiver Solar Thermal Electric System," JOURNAL OF SOLAR ENERGY ENGINEERING, (Paper number 79-WA/Sol-11), Vol. 102, No. 4, pp. 272-280, November, 1980.

10.   Jim Schefter, "Solar power cheaper than coal, oil, gas," POPULAR SCIENCE, Vol. 226, No. 2, pp. 77-79, February 1985.

11.   Graham Walker, "The Stirling Engine," SCIENTIFIC AMERICAN, Vol. 229, No. 2, pp. 80-87, August 1973.

12.   Andy Ross, Stirling Cycle Engines, SOLAR ENGINES (Phoenix, Arizona), p. 31, 1981.

13.   "Annual Technical Report," (DOE/JPL-1060-51) This report was prepared for the U.S. DEPARTMENT OF ENERGY through an agreement with NASA and the JET PROPULSION LABORATORY (Pasadena, California), and NASA LEWIS RESEARCH CENTER (Cleveland, Ohio), March 1982.

14.   George R. Dochat (Mechanical Technology Inc, Latham, New York), "Development of a Small, Free-Piston Stirling Engine, Linear-Alternator System for Solar Thermal Electric Power Applications," SOCIETY OF AUTOMOTIVE ENGINEERS, INC. (SAE), Paper No. 810457, Reference Vol. 90, p. 85, International Congress and Exposition, held in Cobo Hall, Detroit, Michigan, February 23-27, 1981.

15.   E. F. Lindsley, "Solar Stirling Engine," POPULAR SCIENCE, Vol. 212, No. 6, p. 74, June 1978.

16. Glendon M. Benson, Ronald J. Vincent and William D. Rifkin, "An Advanced 15 kW Solar Powered Free-Piston Stirling Engine," 15TH INTERSOCIETY ENERGY CONVERSION ENGINEERING CONFERENCE, (paper number 809-414), pp. 2051-2056, held in Seattle, Washington, August 1980 (Published by the American Institute of Aeronautics and Astronautics, New York, New York).

17. M. A. Liepa and A. Borhan, "High-temperature Steam Electrolysis: Technical and Economic Evaluation of Alternative Process Designs," INTERNATIONAL JOURNAL OF HYDROGEN ENERGY, Vol. 11, No. 7, pp. 435-442, 1986.

18. J. F. Britt, C. W. Schulte and H. L. Davey, "Heliostat Production Evaluation and Cost Analysis," GENERAL MOTORS (GM) TECHNICAL CENTER (Warren, Michigan), December 1979. This report was prepared by GM under sub-contract No. XL-9-8052-1 for the Solar Energy Research Institute (SERI) in Golden, Colorado.

19. Benson, op. cit., p. 2055.

20. Clemens P. Work and Kenneth R. Sheets, "Behind talk of a new wave of oil mergers," U.S. NEWS & WORLD REPORT, Vol. 96, No. 3, p. 59, January 23, 1984.

## CHAPTER 6:
## RENEWABLE ENERGY
## RESOURCES

1. "International Energy Annual 1986," prepared and published by the U. S. ENERGY INFORMATION ADMINISTRATION, (Washington, D.C.), p. 1, October 13, 1987.

2. Ibid., p. 2.

3. Wilson Clark, Energy for Survival, ANCHOR PRESS - DOUBLEDAY (New York, New York), p. 515, 1974.

4. Clarence Zener, "Solar Sea Power," PHYSICS TODAY, Vol. 26, No. 1, p. 52, January 1973.

5.    William H. Avery (Applied Physics Laboratory, Johns Hopkins University), "Ocean Thermal Energy -- Status and Prospects," MTS JOURNAL, Vol. 12, No. 2, p. 52, September 1977.

6.    Zener, op. cit., p. 52.

7.    G. Daniel Brewer (Lockheed Aircraft Company), "Cargo-Carrying Airline proposed for Liquid Hydrogen Fuel Development," ICAO BULLETIN (International Civil Aviation) February 1979. Note: the projected oil reserve data in the graph were compiled by Dr. M. King Hubbert in his paper, "The Energy Resources of the Earth," SCIENTIFIC AMERICAN, Vol. 224, No. 3, pp. 66-70, September 1971.

8.    Arthur F. Pillsbury (former chairman of the Department of Irrigation, University of California at Los Angeles), "The Salinity of Rivers," SCIENTIFIC AMERICAN, Vol. 245, No. 1, pp. 54-65, July 1981.

## CHAPTER 7:
## CONCLUSIONS

1.    Philip Shabecoff, "White House dilutes 'greenhouse' warning," THE NEW YORK TIMES release, published in the ARIZONA REPUBLIC, p. A4, May 8, 1989.

2.    Peter Hoffmann, "Hydrogen gets renewed attention as fuel of the future," THE WASHINGTON POST release, published in THE GRAND RAPIDS PRESS, p. D5, October 11, 1987.

3.    Anne Q. Hoy, "U.S. smog to persist, report says," THE ARIZONA REPUBLIC, pp. A1-A6, July 18, 1989.

4.    Thomas A. Sancton, "What on Earth Are We Doing?" TIME INC. MAGAZINES, Vol. 133, No. 1, p. 30, January 2, 1989.

# Index

**A**

acid deposition 9, 10, 43, 47,
    57, 230, 236
acid fog 10
acid rain 16, 23, 107, 202, 233
Aerobot 145
Age of Exponentials 43
Ages of Gaia: A Biography of
    Our Living Earth 15
AIDS 43
air conditioning 112
Air Force Flight Dynamics
    Laboratory 130
air pollution 9, 23, 107
Air Products and Chemicals
    105
Alaskan oil reserves 68, 235
algae 14
Alkon, Daniel L. 99
Alternative Fuels 148
American Academy of
    Science 121, 124
American Association of
    Physics Teachers 36
American Cancer Society 24
*American Journal of*
    *Physics* 37
American Petroleum
    Institute 24, 45
amino acids 95, 99,102
ammonia 148
Amundson, Robert 23
*Annual Review of Energy* 150
Antarctic 22
Antarctic ozone levels 19

Appleby, John 236
arithmetic savings 40
arithmetic vs. exponential
    growth 37
Arizona State University 210
arms race 28, 29, 30, 49, 201,
    203, 228
Arab oil embargo 58, 72, 125,
    202
Argonne National Laboratory
    83
atmosphere 20, 21, 89
atom 89
"Atoms for Peace" program 203
atomic number 94
atoms 17, 94, 77
atmosphere 22, 23
Avery, W.H. 177, 207

**B**

bacteria 96, 98
bacterial fossils 99
Bartlett, Albert A. 36, 37, 41,
    42, 66, 67, 68, 69, 70, 71
Beale, William T. 188
Benson, Glendon 190, 192
benzene 24
    emissions, 9
Big Bang 94
Billings Energy Corporation
    108, 120, 121, 124
Billings, Roger 121
biohabitat arcology 144

biological transition to
    renewable resources 32
biotechnology 4, 32
Boeing Aircraft Corp. 132
breeder reactors 75
Brewer, G. Daniel 127
Brigham Young University 121
British Antarctic Survey 19
British Interplanetary
    Society 139
British thermal units (Btu) 61,
    197, 198, 206, 209, 218,
    221, 224, 226
Broecker 14
Broecker, Wallace 13
Brookhaven National
    Laboratory 108, 191, 220
Brower, Kenneth and David
    89
Brown, George (U.S. House
    of Representatives) 239
Brown, Harrison 57, 150
Buchner, Helmut 110
budget deficits 57
bulldozer culture 3, 26
Bush, George 73
Bush Administration 234, 235,
    236, 237
Bush, Vannevar 162
Bussard Hydrogen Ramjet
    Spacecraft 140
Bussard, R.W. 139


C

California Institute of
    Technology 10, 88
Canada, energy production 61
cancer 18, 43
cancer statistics 24
carbohydrates 14

carbon 94, 95, 101, 102, 127,
    149
carbon atom 13
carbon cycle 13
carbon dioxide 9, 11, 12, 13,
    14, 15, 16, 25, 26, 43, 57,
    107, 166, 235
Carnegie-Mellon University 103,
    207
Carter Administration 173, 199
Cavendish, Henry 103
CBS 60 Minutes 50
CFC molecules, see
    chlorofluorocarbon
Challenger space shuttle 135
Challenger shuttle explosion
    133
chemical contamination
    problems 24
Chernobyl accident 82
China, energy production 61
chlorine atom 16, 19
chlorofluorocarbon, 9, 11, 12,
    16, 18, 22, 23, 112
citizen action 238
civilization 8
coal 61, 63
    as a primary energy
    source 106
coal-gasification 150
cold fusion (see also nuclear)
    86, 87, 233
Cold War 28, 29
Columbia River 218
Columbia University 13
    Lamont-De-Herty
    Geological Observatory
    177
Commoner, Barry 89
control rods in nuclear fission
    power plants (refer also to
    nuclear) 77

controlled-environment food
production systems 229,
230, 243
Cornell University 23, 46
cosine losses 191
Cosmos 46, 140, 154
crankcase oil 23
crude oil reserves 64
cryogenic 111
cryogenic storage tank weight
111
cryogenic tanker 213

**D**

Daimler-Benz 110
Daimler-Benz, hydrogen
storage systems 108
d'Arsonval 177
Davis, Noel 229-230
decommissioning of nuclear
reactors 78
deforestation 16, 25, 26, 43,
228
Department of Defense
(DOD), hydrogen
research 124
Department of Energy (DOE)
25, 70, 80, 81, 84, 179
desalinated sea water 177, 218-
220
desertification 43
DFVLR (German Aerospace
Research Institute) 118
Diesel cycle engines 185
Diesel engines 186
Diesel, Rudolf 185
dinosaurs 15
dish genset costs 191-194

dish genset systems 62, 157,
161, 178-194, 196-198,
201, 220, 226, 234
deoxyribonucleic acid (DNA) 18,
95, 106
DNA and RNA 100
DNA biocomputer 99
downward mobility 59
driftnetting 166
drought 8, 11, 25, 26

**E**

Earth 36, 45
atmosphere 15, 19, 20,
26, 93
Biosphere 21
climate 15
microbial biota 15
protective ozone shield 18
stratosphere 19
stratospheric ozone
layer 16
economic growth 66
education 49, 58
Einstein, Albert 17, 88
Eisenhower Administration
203
electric waves 18
Electric Power Research
Institute (EPRI) 190
electromagnetic units 17
electrons 7, 13, 17, 89, 94, 126
energy level 17, 18
electric vehicle, battery
storage weight 111
electricity 62
in relationship to
hydrogen 106
electrolysis 125, 218, 220

cost of hydrogen
    production 150
    high-temperature 191,
       220
electromagnetic energy 93
elements, relative abundance of
    102
Energy and Commerce
    Health and
    Environmental
    Subcommittee on
    Ozone Depletion 22
energy crisis 58
*Energy in our Future* 57
energy conservation 31
energy reserves 64
Environmental Protection
    Agency (EPA) 22, 24
enzymes 97
Ericsson, John 184, 185
Erren Rudolf A 119, 120
ethanol 148, 236
ethylene dibromide 24
Evans, Dan (U.S. Senator) 239
evolution of life 17, 96
exponential growth 4, 15, 17,
    25, 26, 33, 35, 36, 37,
    46, 47, 100, 227
    doubling times 38, 40, 42
    positive nature of 45
    rates of change 47
    savings 39
exponential expiration time"
    (EET) 67, 68, 69
Exxon-Alaskan oil spill 235

**F**

fair market 31
Farman 19
Farman, J.C. 19

fermentation 100
Fleischmann, Martin 86
food production systems 229
food shortages 25
food-production systems 27
Ford Motor Company 186, 210
foreign trade deficits 58
*Forever Fuel, The Story of
    Hydrogen* 153
fossil fuels 5,11, 14, 15, 16, 23,
    26, 30, 43, 45, 61, 62,
    71, 73, 75, 88, 90, 93,
    100, 103, 104, 119,
    158, 184, 195, 204,
    205, 218, 228, 236
    liquid fuels 61
    power plants 82, 104
    reserves 66, 74, 84
    world production 60
free market system 31, 173,
    242
free-piston Stirling engines 187-
    189
    electrical generator
    system 189
*Frontline* (PBS) 53
fuel cells, 111, 117
fuel storage tanks 24
Fuller, Buckminster 107
fusion energy systems 178
*Future Shock* 47

**G**

*Gaia:A New Look at Life on
    Earth* 15
Gaian forces 15
gasoline 9, 24, 72, 106, 112,
    118, 119, 127, 150,
    216
    as a liquid hydride 108

compared to liquid
hydrogen 111
storage tank seepage 57
geologic carbon cycle 14
General Dynamics Corp. 212,
213
General Mills Corporation 230
General Motors Corp. 181, 186,
191, 192, 210
German Aerospace
Research
Establishment
(DFVLR) 116
glaciers 25
global acid deposition 10
global solar resources 211
global systems collapse 8
global weather systems 25
Gorbachev, Mikhail 28, 29, 235
Gottingen University 101
governmental regulation 237,
238, 241
Graf Zeppelin 126
gravity 92, 94
green plants 101
greenhouse effect (global
warming) gases 9, 11,
15, 16, 25, 26, 202, 230,
233, 235
Griffith, Jerry 84
Gross National Product (GNP)
66
Grumman Aerospace Corp. 167
Gulf Oil 216


H

Haldane, J.B.S. 162
Hansen, James 12, 235, 236
Harrenstien, Howard 148
Hartz, Jim 50

Hawkins, Willis, M. 152
heliostat 191, 192, 193
helium 93
herbicides 229
Heronemus, William E. 83, 164,
165, 166, 171, 178
Hertz, Heinrich 159
High Frontier 139
high-temperature electrolysis
191, 220
Hindenburg 126, 127, 129
Hinkle, Peter C. 98
Hoffmann, Peter 153
Horgan 85
Horizon 47
Houghton, Richard A. 11, 25
Hubbert, M. King 66, 67, 70,
202
human population growth 41,
43
hydride storage system 110
hydroelectric dams 61, 62, 205,
224
hydrogen (See Chapter 4) 6, 85,
93,170, 181, 190, 196,
206, 216, 218, 220,
238
airport 128
as a chemical feedstock
105, 106
as a primary vs.
secondary energy
source 106
as a universal fuel 104
atom 7, 17
combustion by-products
107
discovery of 103
economy 6, 106, 228
electrolysis 106
energy system 30
explosions 129, 130

flammability 126
from water 217
high-temperature
     electrolysis 191, 220
homestead 121, 122
Hydrides 108
legislation 240
production 149
Production Board 239
reduction of iron or
     copper ores 104
safety 126
storage systems 108
Tappan gas range 121-
     123
hydrogen-fueled biohabitat
     spacecraft 141
Coleman stove 124
engines 119
hypersonic aerospacecraft 135,
     139

**I**

Ice Age 14
icebergs 26
inflation 58, 59, 60, 61
information explosion 46, 48,
     49
*Innovation* (PBS) 50
Integral Fast Reactor 83
interconnections 58
interest rates 58, 72
International Association for
     Hydrogen Energy 6-7,
     153, 190, 199, 216
*International Journal of
     Hydrogen Energy* 153
Interrelationships 57
interstellar ramjet
     "hydrogen scoop" 139

intestinal gases 101
ions 89
Iran 64
Iraq 64
iron-titanium metal hydrides for
     hydrogen storage 124

**J**

Japan 85, 229
     educational system 50
Jet Propulsion Laboratory 88
Johns Hopkins University 207
     Applied Physics
         Laboratory 177

**K**

kerosene 112, 115, 116, 127
     compared to liquid
         hydrogen 111
Khan, Genghis 162
Kilkelly, James 231
kilolangley 209
Kinematic Stirling engines 186
kinetic energy 169
knowledge 45-48
knowledge explosion 45, 46,
     234
Kuwait 64

**L**

La Cour, Poul 162
LaJet Energy Company 182,
     183
Lavi, Abraham 103
Lavoisier, Antoine 103
Law of the Cube 164

life span 45
line-focus solar systems 194-
       195
liquid hydrogen (LH2) 111, 112,
       115, 118, 195, 213,
       214
    compared to hydride
       storage systems 108
    disadvantages 116
    fuel tank 119
    self-service pump 113
liquid hydrogen-fueled aircraft
       115, 116
liquid hydrogen-fueled Buick
       113
liquid sodium 83
Lockheed Aircraft Corp. 116,
       119, 128, 129, 132,
       135, 152, 214, 215
    studies of liquid hydrogen
       fuel tank weight 111
Lockheed Missiles & Space
       Corporation 136, 172,
       173, 174, 175
Los Alamos National
       Laboratory 117, 118
    liquid hydrogen-fueled
       test vehicle 112
    liquid hydrogen self-
       service pump 113
Lovelock, James E. 14
Luz International Limited 194-
       195
lung cancer 22


                M

M-60 tank 125
malignant melanoma 22
mammals 3
Manabe, Syukuro 12
Manhattan Project 7, 49

manned space shuttle launch
       vechicle 133-135
Marchetti, Cesare 100
Mars , carbon dioxide
       assumulations 16
Massachusetts Institute of
       Technology (MIT) 70, 74,
       144, 170, 229
Matsunaga, Spark (U.S.
       Senator) 239
McCarty, Richard E. 98
McDonnell Douglas Corp.136,
       180, 187, 190
Mechanical Technologies Inc.
       (MTI) 189
media 46
metabolism 96
methane 9, 11, 18, 23, 25
methanol 148, 236
    as a liquid hydride 108
Mexico 64
microbes 15, 95, 96
Middle East 30, 64
    oil reserves 71, 200
military industrial complex 203
Ministry of International Trade
       & Industry (MITI) 229,
       242
Mion, Pierre 181
Mobil Oil Corporation 71
Mohave Desert 209
Mohnen, Volker A. 10
molecular mechanisms of
       memory 99
molecular biology 4, 45, 99, 227
molecular chemistry 18
molecular medicine 32, 45
molecules 17, 18, 96
Moller aircraft 145, 146
Moller, Paul 145
Mother Nature 36, 43
Murray, Allen E. 71

**N**

nanometer 18
nanobes (nanoorganisms) 95,
    96, 101, 102
nanobial evolution 96
National Aeronautics & Space
    Administration (NASA)
        19, 22, 130, 132-135,
        179, 186, 189
    wind power study 163
    Goddard Institute for
        Space Studies 11, 235
National Aerospace Plane
    (NASP) 135-139
National debt 43
National Institute of
    Neurological and
    Communicative
    Disorders and Strokes
    99
National Oceanic and
    Atmospheric
    Administration 12
National Sacrifice Zones (for
    nuclear waste) 81
natural gas 62, 125, 148
    as a primary energy
        source 106
neutrons 17, 77
New York University Medical
    Center 22
nickel-59 78
nitrogen 12, 102
nitrous oxides 23
Nixon Administration 199
North American Water and
    Power Alliance
    (NAWAPA) 220, 222-225
nonphotosynthetic bacteria
    226

*Nova* (PBS) science
    documentaries 53
Nuclear Fuel Services (NFS), a
    subsidiary of W.R. Grace's
    Davison Chemical
    Company 78, 79
nuclear power (see also
        radiation) 75, 89, 104,
        125, 158, 218, 235
    as a primary energy
        source 106
    capital costs 163, 201
    cold fusion , 86, 87, 233
    fuel cycle 87
    fission process 62, 77, 82,
        93
    fission breeder reactors
        151
    fusion 85, 93, 126, 151
    economics 87
    high-level nuclear waste
        88
    Integral Fast reactor 83
    melt-down 82
    reactors 77, 184, 190,
        196, 200, 202, 203,
        205
    Three Mile Island 75
    uranium fuel enrichment
        facilities 87
    waste storage 80
nuclear weapons 81
nucleus 89


**O**

Ocean currents, tides and
    waves 157
Ocean thermal energy
    conversion (OTEC)
        systems 62, 157, 171,

172, 173, 175, 176,
    177, 178, 202, 226
  Cutaway View 173
  Operation 174, 175
  resources 207
offshore oil drilling 235
offshore wind energy systems
    165, 166
Ohio University 188
oil
    as a primary energy
        source 106
    prices 58, 61
    reserves 45, 65, 68, 71
    shale 63, 69
    spills 9, 57
    surplus 71
Olson, Arthur J. 97
Omni magazine 89
O'Neill, Gerald, K. 139
Ordin, Paul, M. 130
Othmer, Donald F. 63
Otto, Nikolaus 185
oxides of nitrogen 107
oxygen 12, 14, 16, 102, 141,
    149
ozone 9
    stratospheric 5, 9, 11, 16-
        20, 22, 23, 31, 107,
        112, 227, 230
    tropospheric 20, 21, 23


                P


Parsons Engineering Company
    220, 222
pesticides 11, 229
petroleum (see also oil)
petroleum era in a historical
    perspective 74
    reserves 30

resource 38
pH scale 10
Phoenix bird 6, 95
Phoenix Project 30
photochemical smog 9, 23
photoelectric effect 159
photons 17, 18
photosynthesis 100, 101
photovoltaic cells 62, 157, 158,
        159, 226
    efficiency 161
Phytofarm hydroponic growing
        systems 229-231
Phytofarms of America 229
Pillsbury, Arthur F. 221-224
pipelines 215
plankton 14
plutonium 75, 84
plutonium economy 75, 152
point-focus solar concentrator
    (see also "dish" gensets )
        62, 87, 157, 161, 178-
        194, 197-198, 201
political action 49
Polytechnic Institute of New
        York 63
Pons, Stanley 86
Popular Science  (Times-Mirror
        Magazine) 110, 167-169,
        183, 188
power tower 157, 180, 191-193
President's Office of Science and
        Technology Policy 88
primary vs. secondary energy
        sources 106
primordial soup 100
Princeton University 12, 139
priorities 47, 90
proteins 96, 99
protons 7, 13, 17, 94
Public Broadcast System (PBS)
        50

## Q

quads (quadrillion Btu's) 61, 197, 198, 206, 209, 218, 221, 224, 226

## R

radiation (see also nuclear)
radiation leaks 81
radioactive waste 9, 47, 57, 75, 76, 84, 85, 204, 233
radioactivity 77, 89
rain forests 31
Ralph M. Parsons engineering firm 220, 222
RAND Corporation 70, 74
Ravin, Jack 24
Reagan Administration 72, 73, 173, 187, 200, 201, 237
reforestation 26, 228
Reich, Peter 23
reindustrialization 30, 58, 63
renewable energy 205
Research Institute of Scripps Clinic 97
Return On Investment (ROI) 198
Rickover, Hyman 83
Rigel, Darrel 22
Rockefeller, John D. 242
Ruckelshaus, William D. 32

## S

Sagan, Carl 46, 95, 140, 154
salt accumulation problem 221-225
Sandia National Laboratory 195
Saudi Arabia 64
    energy production 61

Schlegel, Hans 101
*Science* 12
*Scientific American* 10, 12, 25, 32, 85, 96, 98, 99, 221, 223
Sea Solar Power 177
sea water desalination 177, 218-220
Seaborg, Glenn T. 49-50, 52
Senate Bill 639: A Bill For All Interests 239
shale oil (see oil)
Shaw, Robert W. 10
skin cancer 22
Smith-Putnam wind generator 163
solar energy (see Chapters 5 & 6) 62, 100, 102, 103, 105, 151, 206
    dish forest 181
    dish genset systems 62, 157, 161, 178-194, 196-198, 201, 220, 226, 234
    economics 196
    radiation 17-19
    technologies 104, 157-158, 204, 228
    thermal systems 157
solar-hydrogen 90, 228
SolarPlant-1 183
Soleri, Paolo, 144, 146, 170
solutions 228
Sonoran Desert 209
Southern California Edison Company 180,187, 190, 193-194
Soviet Union 64, 82, 85, 88, 200
    arms race 28-30, 174, 199-201, 228
    energy production 61

oil production 72
solar resources 212
Space Habitats 139-147
Space Shuttle 132-139
specialization 58
Starship Hydrogen 141, 147
Starship Hydrogen Core
  Structure 142
State University of New York 10
stellar hydrogen 92
Sternback, Rick 140
Stirling "sun motor" 184-185
Stirling dish gensets for large-
  scale hydrogen production
  190
Stirling engine 180, 184-190,
  192
Stirling, Robert 184
stratospheric ozone depletion,
  stratospheric 5, 9, 11, 16-
  20, 22, 23, 48, 31,
  107, 112, 223, 226
  the impact of hydrogen as
  a fuel 107
stratospheric ozone layer 17,
  19, 20, 23, 31
strip-mining 57
stripmining the seas via
  driftnetting practices 167
supersonic transport (SST)
  program 137
sunlight 22
superorganism 14
systems collapse 44

T

Tappan Appliance Company
  109, 122
  hydrogen-fueled Gas
  Range 123

tar sands 63
teachers 50, 52
Texas Tech University 60
The Closing Circle 89
The New York Times 79
Third World countries 58
Thomas, Percy H. 162
Thorndike, Edward H. 58
Three Mile Island (TMI) nuclear
  power facility 75, 131
Till, Charles E. 83
Titanic 35, 36, 55, 75, 235-236
Toffler, Alvin 47, 48
top soil 25
toxic chemicals 43
transhydrogenation 100
tropospheric ozone 23, 57
Texas A&M's Center for
  Electrochemical Systems
  and Hydrogen
  Research 236
topsoil erosion 230
toxic chemicals 3, 228

U

ultraviolet radiation 16, 17, 18,
  19
underground cities 230
United States.(U.S.) 22, 24, 36,
  88
  coal reserves 62
  Congress 80, 232, 239
  crude oil reserves 64
  Department of Defense
  (DOD) 124, 130
  Department of Energy
  (DOE) 25, 70, 80, 81,
  84, 179
  dollar 59, 179
  economy 66

educational system 49
Energy Information
    Administration 61, 63,
    65, 76, 205
energy policy 199
energy production 61, 205
Environmental
    Protection Agency
    (EPA) 22, 24
fusion research 85
Geological Survey (USGS)
    45, 70, 74, 81, 202
grain production 8
Naval Research
    Laboratory 181
Navy nuclear submarines
    83
Office of Technology
    Assessment 236
oil production 71
oil reserves 71
solar resources 210
vs. Soviet military forces
    28
unemployment 57
United Kingdom energy
    production 61
United Stirling of Sweden 180
universe 3, 92
University of California at
    Davis 145
University of California at Los
    Angeles (UCLA) 10, 49,
    221
University of Colorado 36, 66
University of Massachusetts at
    Amherst 83, 165
University of Miami 148, 153
University of Rochester 58
University of Southampton 86
University of Utah 86, 233
uranium-239 79, 204

fuel rods 77
    reserves 84


V

V-2 rockets 132
Venus 16
Veziroglu, T. Nejat (see also
    International Association
    for Hydrogen Energy) 153
Von Braun, Warner 132
Vertical Vortex Wind
    Generators 167-168-170


W

water considerations 205-208,
    216-226
    desalination 177, 218-220
    options 218
    for hydrogen production
        216-223
    vapor 206
    for energy production 208
    North American Water
        and Power Alliance
        (NAWAPA) 220-225
    oil spills 214
Watson, Robert 22
Wavelength 17
Wavelength and Energy Level
    18
weather records 25
West Valley nuclear fuel
    reprocessing plant 78, 79
Westinghouse Electric Corp.
    191, 220
Wetherald, Richard 12
wilderness areas 57
wildlife habitats 43

Wilson, Carroll 70

wind energy conversion systems
        62, 157, 161-171, 202,
        207, 226
    disadvantages 163
    environmental emissions
        166
    Law of the Cube 164
    energy resource 206
    energy production 205
    vertical vortex systems
        167-170

Woese, Carl R. 96

Woods Hole Research Center
        11, 12

Woodwell 12, 25

Woodwell, George M. 11

*World Watch* magazine 8

*World* Magazine 107

world
    coal production 63
    crude oil production 65
    energy production 205
    nuclear production 76
    oil production 215
    oil reserves 70

World Meteorological
    Organization (wind energy
    data) 206


**Y**

Yen, James T. 167, 169

Yucca Mountain nuclear waste
    storage site 80


**Z**

Zener, Clarence 103, 207

**HARRY W. BRAUN**
Research Analyst

Harry W. Braun is a research analyst specializing in energy and environmental research. He has been specifically involved in a comprehensive analysis of energy technologies, resources, and economics since 1974. As a result of this comprehensive research effort, he has been able to interact with key engineers and scientists from a wide range of industrial and governmental facilities, including Ford Aerospace Corporation, Lockheed Aircraft Corporation, General Dynamics Corporation, Grumman Aerospace Corporation, Garrett Turbine Engine Company, General Motors Corporation, National Aeronautics and Space Administration (NASA), Department of Energy, Los Alamos National Laboratory and the Jet Propulsion Laboratory.

From 1982 to the present, Braun has been working as a research analyst with Trans Energy Corporation (Phoenix, Arizona), an engineering and materials science firm that is involved in developing state-of-the-art and advanced automotive, energy and environmental control systems. Braun is a member of the American Society for Photobiology (McLean, Virginia), and he is an advisory board member of the International Association for Hydrogen Energy (Coral Gables, Florida) and the American Hydrogen Association (Tempe, Arizona). He is a graduate of Arizona State University (Tempe, Arizona), and is a former science teacher in the Glendale Union High School District (Glendale, Arizona).

\*   \*   \*